21世纪高等学校计算机类课程创新规划教材·微课版

Android Studio
应用程序设计（第2版）

◎ 张思民　编著

微课版

清华大学出版社
北京

内 容 简 介

本书是面向 Android Studio 初学者的入门教程，内容基本涵盖了 Android 相关的所有技术，大致可以分成两个部分。第一部分（第 1～4 章）主要介绍 Android Studio 的安装、应用程序的结构、图形用户界面的组件及其设计方法；第二部分（第 5～9 章）主要介绍较高级的主题，内容包括后台服务与系统服务、网络通信技术、数据存储技术、地图服务与传感器检测技术等。

本书内容浅显易懂，对每一个知识点都配了相应的例题。本书提供了电子课件和所有例题的源代码，扫描每章提供的二维码可观看教学视频。

本书可作为高等院校及各类培训学校 Android 系统课程的教材，也可作为希望学习 Android 系统开发的读者的自学用书。

本书封面贴有清华大学出版社防伪标签，无标签者不得销售。
版权所有，侵权必究。侵权举报电话：010-62782989　13701121933

图书在版编目（CIP）数据

Android Studio 应用程序设计：微课版/张思民编著. —2 版. —北京：清华大学出版社，2017（2019.12重印）
（21 世纪高等学校计算机类课程创新规划教材·微课版）
ISBN 978-7-302-48134-8

Ⅰ. ①A…　Ⅱ. ①张…　Ⅲ. ①移动终端 -应用程序 -程序设计　Ⅳ. ①TN929.53

中国版本图书馆 CIP 数据核字（2017）第 203703 号

责任编辑：魏江江　王冰飞
封面设计：刘　键
责任校对：胡伟民
责任印制：刘海龙

出版发行：清华大学出版社
网　　址：http://www.tup.com.cn, http://www.wqbook.com
地　　址：北京清华大学学研大厦 A 座　　邮　编：100084
社 总 机：010-62770175　　邮　购：010-62786544
投稿与读者服务：010-62776969，c-service@tup.tsinghua.edu.cn
质 量 反 馈：010-62772015，zhiliang@tup.tsinghua.edu.cn
课 件 下 载：http://www.tup.com.cn, 010-62795954

印 装 者：清华大学印刷厂
经　　销：全国新华书店
开　　本：185mm×260mm　　印　张：17.75　　字　数：435 千字
版　　次：2013 年 3 月第 1 版　　2017 年 10 月第 2 版　　印　次：2019 年 12 月第 7 次印刷
印　　数：23001～26500
定　　价：39.50 元

产品编号：075773-01

第2版前言

与第 1 版相比，第 2 版的最大修改之处是把开发工具 Eclipse 换成了 Android Studio。这主要考虑 Android Studio 是 Google 公司推出的专门为 Android "量身定制"的开发工具，是 Google 大力支持的一款基于 IntelliJ IDEA 改造的 IDE 集成开发环境，可以说它是 Android 开发工具的未来。

第 2 版删除了第 1 版中 Java 语言基础知识简介的章节，增加了访问 Web 服务器数据的内容。另外，由于 Android Studio 与 Eclipse 的目录结构和菜单差异很大，因此很多例题在介绍操作时做了修改。

第 2 版全书共分 9 章，第 1 章主要讲解 Android Studio 开发环境的搭建，并介绍了开发 Android 应用程序的步骤和应用程序框架的结构；第 2~3 章讲解如何使用布局和视图创建用户界面，介绍了用户图形界面的常用组件；第 4 章介绍图形与多媒体处理技术，介绍了绘制几何图形的基本方法、处理触摸屏事件的方法，还详细讨论了音频播放和视频播放的设计以及录音、照相技术，并详细讲解了在 Android 中实现动画的技术；第 5 章介绍后台服务与系统服务，以及系统功能调用；第 6 章介绍网络通信技术，介绍了 Web 视图以及基于 TCP 协议的网络程序设计、基于 HTTP 协议的网络程序设计等网络编程技术；第 7 章介绍应用 Volley 框架访问 Web 服务器，并介绍了 JSON 数据格式和一个网络音乐播放器设计实例；第 8 章介绍数据存储技术，介绍了 SQLite 数据库存储方式、文件存储方式和 XML 文件的 SharedPreferences 存储方式，还介绍了访问远程数据库的方法；第 9 章讲解地图服务与传感器检测技术，地图服务主要介绍实现地图视图的基本方法，传感器检测主要介绍重力加速度的应用。

本书提供了电子课件和所有例题的源代码，扫描每章提供的二维码可观看教学视频。

张思民
2017 年 5 月

目　录

第1章　Android 系统及其开发过程 ··· 1
　1.1　Android 系统概述 ··· 1
　　　1.1.1　Android 系统及特点 ··· 1
　　　1.1.2　Android 系统的体系结构 ······································ 2
　　　1.1.3　Android 开发的分类 ··· 3
　1.2　搭建 Android Studio 开发环境 ··· 4
　　　1.2.1　安装 Android Studio 前的必要准备 ······················· 4
　　　1.2.2　安装 Android Studio 详解 ······································ 5
　　　1.2.3　设置环境变量 ··· 7
　1.3　Android API 和在线帮助文档 ·· 8
　1.4　Android 应用程序的开发过程 ··· 9
　　　1.4.1　开发 Android 应用程序的一般过程 ····················· 9
　　　1.4.2　生成 Android 应用程序框架 ································ 10
　　　1.4.3　编写代码生成 MainActivity.java ·························· 12
　　　1.4.4　在模拟器中运行应用程序 ··································· 13
　1.5　Android 项目结构 ·· 13
　　　1.5.1　目录结构 ··· 13
　　　1.5.2　Android 应用程序结构分析 ································· 19
　1.6　Android 应用程序设计示例 ··· 21
　习题 1 ·· 23

第2章　Android 用户界面的设计 ·· 24
　2.1　用户界面组件包 widget 和 View 类 ······························· 24
　2.2　文本标签 TextView 与按钮 Button ································· 25
　　　2.2.1　文本标签 ··· 25
　　　2.2.2　按钮及按钮处理事件 ·· 27
　2.3　文本编辑框 ··· 32
　2.4　Android 布局管理 ·· 35
　　　2.4.1　布局文件的规范与重要属性 ······························· 36
　　　2.4.2　常见的布局方式 ·· 37
　2.5　进度条和选项按钮 ··· 46

    2.5.1 进度条 46
    2.5.2 选项按钮 48
  2.6 图像显示类 ImageView 与画廊组件类 Gallery 55
    2.6.1 图像显示类 ImageView 55
    2.6.2 画廊组件类 Gallery 与图片切换器 ImageSwitcher 59
  2.7 消息提示类 Toast 64
  2.8 列表组件 67
    2.8.1 列表组件类 ListView 67
    2.8.2 ListActivity 类 69
  2.9 滑动抽屉组件类 SlidingDraw 72
  习题 2 76

第 3 章 多个用户界面的程序设计 78

  3.1 页面的切换与传递参数值 78
    3.1.1 传递参数组件 Intent 78
    3.1.2 Activity 页面的切换 78
    3.1.3 应用 Intent 在 Activity 页面之间传递数据 82
  3.2 菜单设计 86
    3.2.1 选项菜单 86
    3.2.2 上下文菜单 89
  3.3 对话框 91
    3.3.1 消息对话框 AlertDialog 91
    3.3.2 其他几种常用对话框 96
  习题 3 100

第 4 章 图形与多媒体处理 101

  4.1 绘制几何图形 101
    4.1.1 几何图形绘制类 101
    4.1.2 几何图形的绘制过程 102
    4.1.3 自定义组件 107
  4.2 触摸屏事件的处理 110
    4.2.1 简单的触摸屏事件 110
    4.2.2 手势识别 116
  4.3 音频播放 120
    4.3.1 多媒体处理包 120
    4.3.2 多媒体处理播放器 MediaPlayer 120
    4.3.3 播放音频文件 121
  4.4 视频播放 127
    4.4.1 应用媒体播放器播放视频 127

4.4.2　应用视频视图播放视频 130
4.5　录音与拍照 133
　　4.5.1　用于录音、录像的 MediaRecorder 类 133
　　4.5.2　录音示例 134
　　4.5.3　拍照 137
4.6　动画技术 142
　　4.6.1　动画组件类 142
　　4.6.2　补间动画 Tween Animation 143
　　4.6.3　属性动画 Property Animation 148
习题 4 152

第 5 章　后台服务与系统服务 153
5.1　后台服务 Service 153
5.2　信息广播机制 Broadcast 157
5.3　系统服务 166
　　5.3.1　Android 的系统服务 166
　　5.3.2　系统通知服务 Notification 167
　　5.3.3　系统定时服务 AlarmManager 169
　　5.3.4　系统功能的调用 172
习题 5 175

第 6 章　网络通信技术 176
6.1　Web 视图 176
　　6.1.1　浏览器引擎 WebKit 176
　　6.1.2　Web 视图对象 176
　　6.1.3　调用 JavaScript 180
6.2　基于 TCP 协议的网络程序设计 186
　　6.2.1　网络编程的基础知识 187
　　6.2.2　利用 Socket 设计客户机/服务器系统程序 191
　　6.2.3　应用 Callable 接口实现多线程 Socket 编程 196
6.3　基于 HTTP 协议网络程序设计 200
　　6.3.1　建立 PHP 服务器网站 200
　　6.3.2　应用 HttpURLConnection 访问 Web 服务器 200
习题 6 210

第 7 章　应用 Volley 框架访问 Web 服务器 211
7.1　Volley 框架及其应用 211
　　7.1.1　Volley 包的下载与安装 211
　　7.1.2　JSON 数据格式简介 212

     7.1.3　Volley 的工作原理和几个重要对象 ································· 216
     7.1.4　Volley 的基本使用方法 ·· 217
  7.2　应用 Volley 框架设计网络音乐播放器 ······································ 221
  习题 7 ··· 226

第 8 章　数据存储技术 ·· 227

  8.1　SQLite 数据库 ··· 227
     8.1.1　SQLite 数据库简介 ·· 227
     8.1.2　管理和操作 SQLite 数据库的对象 ································ 228
     8.1.3　SQLite 数据库的操作命令 ··· 228
  8.2　文件的处理 ··· 240
     8.2.1　输入/输出流 ·· 240
     8.2.2　处理文件流 ··· 241
  8.3　轻量级存储 SharedPreferences ·· 248
  8.4　访问远程数据库 ·· 250
  习题 8 ··· 255

第 9 章　地图服务与传感器检测技术 ·· 256

  9.1　电子地图服务的应用程序开发 ·· 256
     9.1.1　Android 地图的 SDK 开发包的下载以及 Key 的申请 ············ 256
     9.1.2　显示地图的应用程序示例 ·· 259
  9.2　传感器检测技术 ·· 262
     9.2.1　传感器简介 ··· 262
     9.2.2　加速度传感器的应用示例 ·· 264
  习题 9 ··· 273

第1章 Android 系统及其开发过程

1.1 Android 系统概述

1.1.1 Android 系统及特点

2007 年 11 月 5 日，Google 公司推出了基于 Linux 操作系统的智能手机平台 Android 系统。Android 系统由操作系统、中间件、用户界面程序和应用软件等组成。2013 年 5 月 16 日，Google 公司推出新的 Android 开发环境——Android Studio。Android 的出现绝非偶然，是由传统的移动电话系统开发模式演变而来的一种符合时代潮流的新型移动开发模式的产物。

Android 传奇的创造与被称为 Android 之父的 Android 创始人安迪·鲁宾（Andy Rubin）密不可分。2003 年，安迪·鲁宾成立了一家叫 Android 的公司，致力于开发一个面向所有软件设计者开放的移动手机平台。安迪·鲁宾的 Android 项目因顺利开展受到了一些风险投资公司的青睐。2005 年 3 月，由安迪·鲁宾继续负责 Android 项目的研发工作。2007 年 11 月，Google 公司正式公布了 Android 操作系统，并且宣布与 34 家手机厂商、运营商成立"开放手机联盟（OHA）"，自此这个基于 Linux 内核的 Android 系统正式登上历史舞台。在接下来的 5 年里，安迪·鲁宾负责的 Android 系统获得了令人难以置信的成功。到 2013 年，在市场占有率方面，Google 公司的 Android 系统主导了手机世界。

Android 系统诞生在开放时代的背景下，其全开放的智能移动平台、多硬件平台的支持、使用众多标准化的技术、完整的核心技术、完善的 SDK 和文档、完善的辅助开发工具等特点与智能手机的发展方向紧密相连，它将代表并引领新时代的技术潮流。

Android 系统具有开放性、平等性、方便性及硬件丰富性等特点。下面对这些特点进行简单介绍。

1. 系统开放性

Android 系统是一款真正开放的系统。Android 系统从底层的操作系统直到最上层的应用程序都是开放的，程序开发人员和爱好者都可以很方便地从网络上获取到源代码，可以对它们进行分析和移植。

2. 应用程序平等性

在 Android 系统开发平台上 Android 系统自带的程序与程序开发人员自己开发的应用程序都是平等的，程序开发人员可以开发个人喜爱的应用程序来代替系统的程序，构建个性化的 Android 系统。

3. 开发方便性

在 Android 系统开发平台上开发应用程序是非常方便的，Android 系统为开发人员提供了大量的实用组件库和方便的工具，开发人员只需编写几行代码就可以将功能强大的组件添加到自己的程序中。

4. 硬件丰富性

由于 Android 系统的开放性，众多的硬件制造商纷纷开发出各种各样的可以与 Android 系统兼容的产品，进一步丰富了 Android 系统的应用。

1.1.2 Android 系统的体系结构

Android 系统的体系结构和其操作系统一样，采用了分层的架构。Android 系统分为 4 层，从顶层到底层分别是应用程序层、应用程序框架层、系统运行库层和 Linux 核心层，如图 1.1 所示。

1. 应用程序

Android 系统自带了一套核心应用程序，包括电话拨号程序、短信程序、日历、音乐播放器、浏览器、联系人管理程序等，如图 1.2 所示。所有的应用程序都是用 Java 语言编写的，开发人员自己开发的应用程序就位于应用程序层。该层的程序是完全平等的，开发人员可以任意用自己开发的应用程序替换 Android 系统自带的程序。

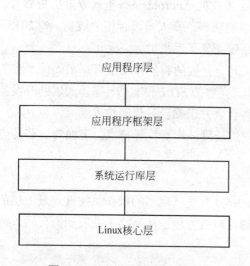

图 1.1　Android 系统的体系结构

图 1.2　Android 系统自带的应用程序

2. 应用程序框架

Android 系统通过应用程序框架为开发人员创建自己的应用程序提供了一个开放的开

发平台，程序开发人员可以在这个应用程序框架平台上设计自己的应用程序。本书所讲的程序设计都是基于这个应用程序框架完成设计的。

Android 系统的应用程序框架主要包含以下 9 个部分。
- 活动页面管理（Activity Manager）：用于管理程序的生命周期。
- 窗口管理（Window Manager）：用于管理应用程序窗口。
- 内容供应（Content Providers）：提供数据共享，使一个应用程序可以访问另一个应用程序的数据。
- 视图系统（View System）：用于构建应用程序的可视化组件。
- 包管理（Package Manager）：用于管理项目程序。
- 电话管理（Telephone Manager）：移动设备的基本功能统一由电话管理器管理。
- 资源管理（Resource Manager）：为应用程序提供所需的文字、声音、图片、视频或布局文件等资源。
- 位置管理（Location Manager）：用于提供位置服务。
- 通知管理（Notification Manager）：在手机顶部的状态栏中发布消息提示。

3. 系统运行库

1）程序库

Android 包含一些 C/C++程序库，这些库能被 Android 系统中不同的组件使用。它们通过 Android 应用程序框架为开发者提供服务。

2）Android 运行时库

Android 包含一个核心库，该核心库提供了 Java 编程语言核心库的大多数功能。Android 系统的 Dalvik 虚拟机也包含在 Android 运行时库中。

4. Linux 内核

Android 的核心系统服务依赖于 Linux 内核，其安全性、内存管理、进程管理、网络协议栈和驱动模型等基本依赖于 Linux。

1.1.3 Android 开发的分类

对于开发者而言，Android 开发分为以下两大类。

1. 系统移植开发

移植开发是为了使 Android 系统能在手持式移动设备上运行，在具体的硬件系统上构建 Android 软件系统。这种类型的开发在 Android 底层进行，需要移植开发 Linux 中相关的设备驱动程序及 Android 本地框架中的硬件抽象层，也就是需要将设备驱动与 Android 系统联系起来。Android 系统对硬件抽象层都有标准的接口定义，在移植时只需实现这些接口即可。

2. Android 应用程序开发

应用程序开发可以基于硬件设备(用于测试的实体手机)，也可以基于 Android 模拟器。应用程序开发处于 Android 系统的顶层，使用 Android 系统提供的 Java 框架（API）进行开发设计工作，是大多数开发者从事的开发工作。本书所介绍的 Android 应用程序设计都是在这个层上进行的。

1.2 搭建 Android Studio 开发环境

1.2.1 安装 Android Studio 前的必要准备

1. Android 系统开发的操作平台

Android 系统开发的软件环境目前有两种，一种是 Eclipse + ADT（Android Development Tools 插件）系统，另一种是 Android Studio 系统。在这里主要介绍 Android Studio 系统。

Android Studio 是一个全新的基于 IntelliJ IDEA 的 Android 开发环境（IntelliJ IDEA 是一种用 Java 语言开发的集成开发环境，是被业界公认为最好的 Java 开发工具），Android Studio 提供了集成的 Android 开发工具用于应用程序的开发和调试。

在安装 Android Studio 之前需要安装 Java JDK 的环境。

2. 下载最新版本的 Android Studio 软件

读者可以到 Android Studio 官方网站"http://developer.android.com/sdk/index.html"免费下载最新的系统软件，如图 1.3 所示。

图 1.3　Android Studio 官方下载页面

进入下载页面以后下载对应操作系统所支持的版本，见表 1-1（以 Android Studio 2.1 版本为例）。

表 1-1　下载 Android Studio 系统安装包

安装平台	系统安装包	Size
Windows	android-studio-bundle-143.2739321-windows.exe 包含 Android SDK（推荐）	1166 MB
	android-studio-bundle-143.2739321-windows.exe	264 MB
Mac OS X	android-studio-ide-143.2739321-mac.dmg	279 MB
Linux	android-studio-ide-143.2739321-linux.zip	278 MB

1.2.2　安装 Android Studio 详解

1. 按安装向导完成 Android Studio 系统的安装

运行安装文件 android-studio-bundle-143.2739321-windows.exe，按照安装向导完成系统的安装，如图 1.4 所示。

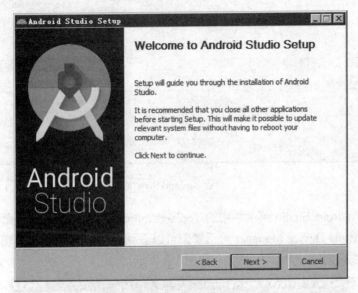

图 1.4　Android Studio 系统的安装

2. 设置 Android SDK 的存放位置

安装完成后，第一次运行 Android Studio 系统需要设置 Android SDK，找到 SDK 的存放位置，如图 1.5 所示。Android SDK 的存放位置也可以通过 Android Studio 应用程序的 Settings 命令设置。

3. 创建 Android 虚拟设备 AVD

Android 应用程序可以在实体手机上执行，也可以创建一个 Android 虚拟设备 AVD（Android Virtual Device）来测试。每一个 Android 虚拟设备 AVD 模拟一套虚拟环境来运行 Android 操作系统平台，这个平台有自己的内核、系统图像、外观显示、用户数据区和仿真的 SD 卡等。

下面介绍如何创建一个 Android 虚拟设备 AVD。

Android Studio 集成开发环境提供了 Android Virtual Device Manager 功能，用户可以用它来创建和调用 Android 虚拟设备 AVD。

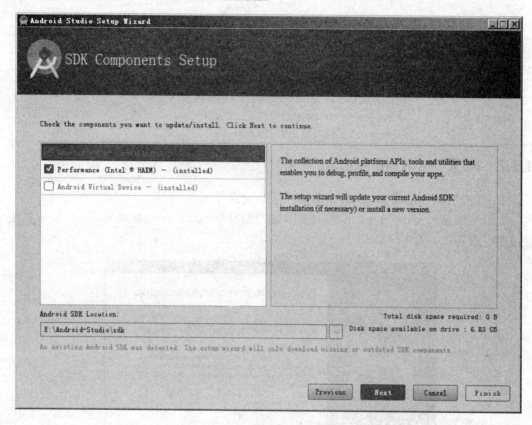

图 1.5 设置 Android SDK 的位置

（1）选择 Android Studio 菜单栏中的 Tools→Android→AVD Manager 命令，用户在弹出的 Android Virtual Device Manager 对话框中可以看到已创建的 AVD。单击下方的 Create Virtual Device 按钮创建一个新的 AVD，如图 1.6 所示。

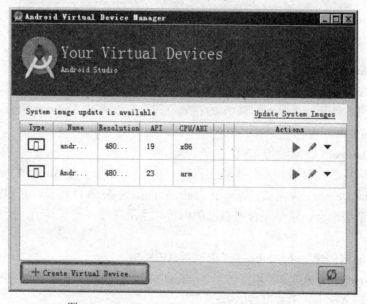

图 1.6 Android Virtual Device Manager 对话框

（2）运行 AVD 模拟器。在 Android Virtual Device Manager 对话框中选择已经建立的 AVD，单击 Actions 栏中的 ▶ 按钮可以启动 AVD 模拟器。启动 AVD 模拟器的时间很长，建议打开后不要关闭，可以在该模拟器上测试 Android 应用程序。启动的 AVD 模拟器如图 1.7 所示。

图 1.7 Android 的 AVD 模拟器

1.2.3 设置环境变量

安装完 Android Studio 系统后还要设置环境变量，即把 Android Studio 系统目录下的 platform-tools 路径设置到系统变量中。右击桌面上的"计算机"图标，在弹出的快捷菜单中选择"属性"命令，在"控制面板\系统"对话框中选择"高级系统设置"选项，再单击"环境变量"按钮，在打开的"环境变量"对话框的"系统变量"下方找到 Path 变量，单击"编辑"按钮，在"编辑系统变量"对话框的"变量值"栏中输入 Android Studio 安装目录下的 platform-tools 完整路径和 tools 完成路径，如图 1.8 所示。

例如设安装路径如下：

C:\Users\Administrator\AppData\Local\Android\sdk

则需要增加设置 Path 变量的值：

C:\Users\Administrator\AppData\Local\Android\sdk\tools

以及

C:\Users\Administrator\AppData\Local\Android\sdk\ platform-tools

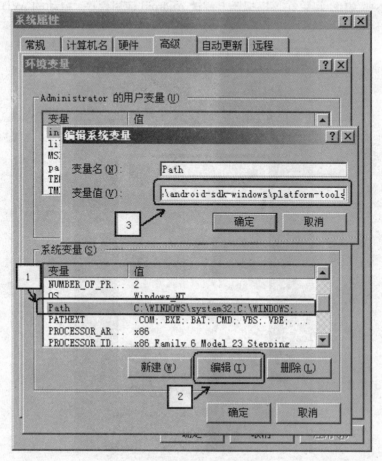

图1.8 设置Android环境变量

1.3 Android API 和在线帮助文档

1. Android API

Android 为用户安装了它所提供的标准类库。所谓标准类库，就是把程序设计所需要的常用方法和接口分类封装成包，Android 提供的标准类库就是 Android API。

在 Android 包中封装了程序设计所需要的主要应用类，本书中所用到的包如下。

- Android.app：封装了顶层的程序模型，提供基本的运行环境。
- Android.content：封装了各种对设备上的数据进行访问和发布的类。
- Android.database：通过内容提供者浏览和操作数据库。
- Android.graphics：底层的图形库，包含画布、颜色过滤、点、矩形，可以将它们直接绘制到屏幕上。
- Android.location：封装了定位和相关服务的类。
- Android.media：封装了一些类管理多种音频、视频的媒体接口。

- Android.net：封装了帮助网络访问的类，超过通常的 java.net.* 接口。
- Android.os：封装了系统服务、消息传输、IPC 机制。
- Android.opengl：封装了 opengl 的工具、3D 加速。
- Android.provider：封装了类访问 Android 的内容提供者。
- Android.telephony：封装了与拨打电话相关的 API 交互。
- Android.view：封装了基础的用户界面接口框架。
- Android.util：涉及工具性的方法，例如时间、日期的操作。
- Android.webkit：默认浏览器操作接口。
- Android.widget：封装了各种 UI 元素（大部分是可见的）在应用程序的屏幕中使用。

2. Android API 帮助文档

Android Studio 提供了离线的 Android API 文档，这是进行程序设计的好工具，希望大家都能用好这个工具。

运行 Android Studio 安装目录下的 index.html 文件，运行结果如图 1.9 所示。

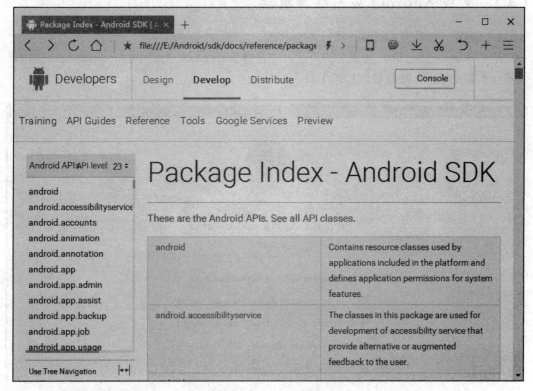

图 1.9　Android API 在线帮助文档

1.4　Android 应用程序的开发过程

1.4.1　开发 Android 应用程序的一般过程

开发 Android 应用程序的一般过程如图 1.10 所示。

图 1.10 Android 应用程序的开发过程

1.4.2 生成 Android 应用程序框架

1. 创建一个新的 Android 项目

启动 Android Studio，在弹出的对话框中选择 Start a new Android Studio project 选项，创建一个新的 Android 项目，如图 1.11 所示。

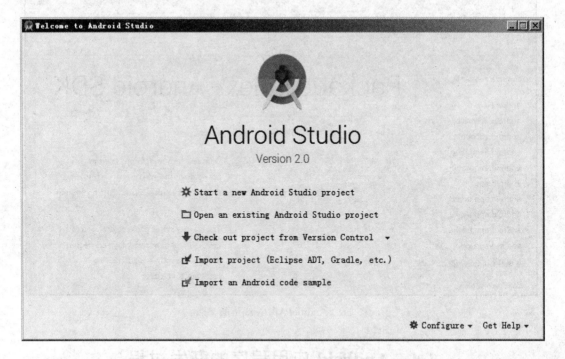

图 1.11 新建一个 Android 项目

2. 填写应用程序的参数

在弹出的对话框中按步骤输入应用程序名称、项目名称、包名等参数，并选择 Android Studio 的版本，如图 1.12 和图 1.13 所示。

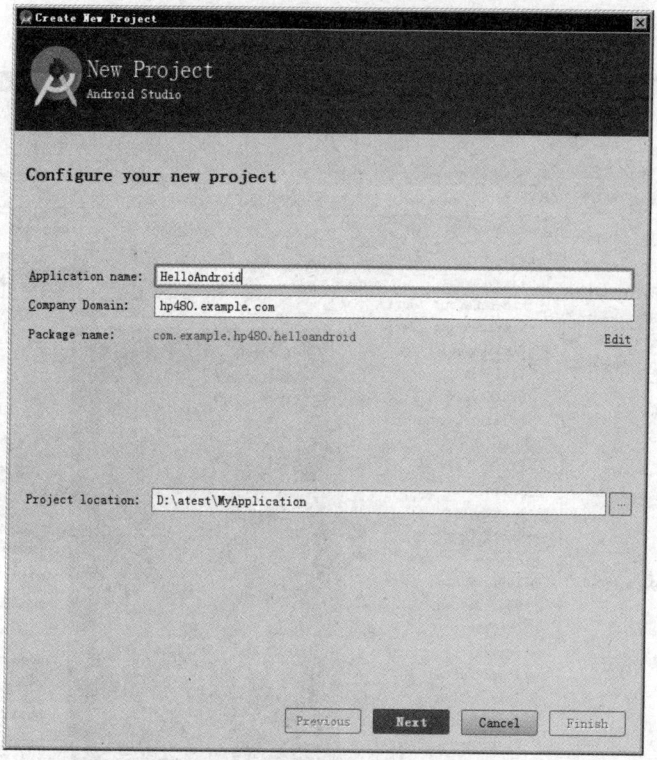

图 1.12 输入 Android 项目名称

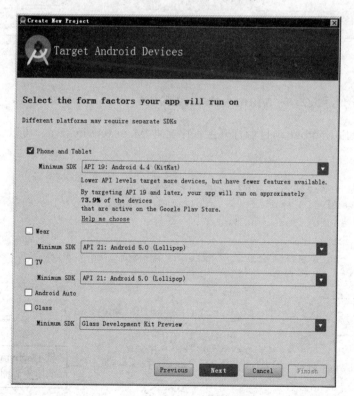

图 1.13 选择 Android SDK 的版本

最后系统自动生成一个 Android 应用项目框架，如图 1.14 所示。

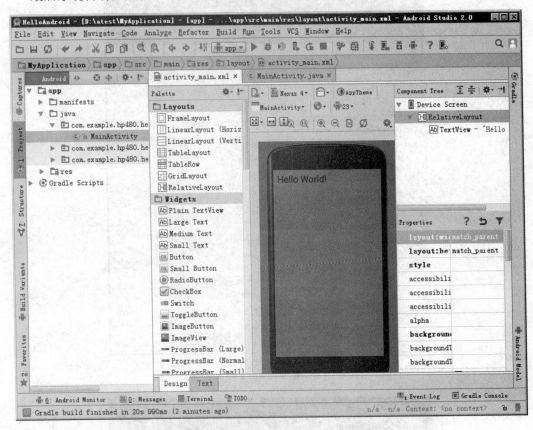

图 1.14　系统自动生成的 HelloAndroid 应用项目框架

1.4.3　编写代码生成 MainActivity.java

在创建 HelloAndroid 项目后打开主程序文件 MainActivity.java，可以看到系统自动生成的代码如下：

```
1 package com.example.hp480.helloandroid;
2 import android.support.v7.app.AppCompatActivity;
3 import android.os.Bundle;

4 public class MainActivity extends AppCompatActivity
5 {
6    @Override
7    public void onCreate(Bundle savedInstanceState)
8    {
9        super.onCreate(savedInstanceState);
10       setContentView(R.layout.activity_main);   ← 显示 activity_main.xml 定义的用户界面
11   }
12 }
```

在 Android 系统中应用程序的入口程序（主程序）都是活动程序界面 Activity 类的子类。在上述代码中最重要的是第 10 行，用于显示用户界面。

1.4.4 在模拟器中运行应用程序

在工具栏中单击 Run App 按钮 ▶ 运行 AVD 模拟器，可以看到应用程序的运行结果（首次运行程序可能耗时较长），如图 1.15 所示。

图 1.15 在 AVD 模拟器设备上显示程序运行结果

1.5 Android 项目结构

1.5.1 目录结构

打开 HelloAndroid 项目，在项目资源管理器中可以看到应用项目的目录和文件结构，如图 1.16 所示。

实际上，图 1.16 所示的目录结构内容是最基本的，程序员还可以在此基础上添加需要的内容。下面对 app 模块下的文件目录结构的基本内容进行介绍。

（1）manifests：其下的 AndroidManifest.xml 为项目的配置信息文件。

（2）java：主要是源代码和测试代码。

（3）res：主要是资源目录，存储所有的项目资源。

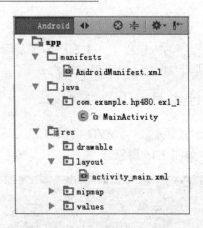

图 1.16　HelloAndroid 项目的目录和文件结构

- values：存储 app 引用的一些值。
  - colors.xml：存储了一些 color 的样式。
  - dimens.xml：存储了一些公用的 dp 值。
  - strings.xml：存储了引用的 string 值。
  - styles.xml：存储了 app 需要用到的一些样式。
- Gradle Scripts:build.gradle 为项目的 gradle 配置文件。

1. java 目录

java 目录存放 Android 应用程序的 Java 源代码文件。在系统自动生成的项目结构中有一个在创建项目时输入 Create Activity 名称的 Java 文件 MainActivity.java，如图 1.17 所示。

图 1.17　src 目录下的 MainActivity.java 的源代码

2. 资源目录 res 及资源类型

Android 系统的资源为应用项目所需要的声音、图片、视频、用户界面文档等，其资源文件存放于项目的 res 目录下。资源的目录结构及类型如表 1-2 所示。

表 1-2　Android 系统的资源目录结构及类型

目录结构	资源类型
res\values	存放字符串、颜色、尺寸、数组、主题、类型等资源
res\layout	XML 布局文件

续表

目录结构	资源类型
res\drawable	图片（bmp、png、gif、jpg等）
res\anim	XML格式的动画资源（帧动画和补间动画）
res\mipmap	存储系统的图片资源
res\raw	可以存放任意类型的文件，一般存放比较大的音频、视频、图片或文档，会在R类中生成资源id，封装在apk中
assets	可以存放任意类型，不会被编译，与raw相比，不会在R类中生成资源id

（1）目录mipmap存储系统的图片资源，*dpi表示存储不同分辨率的图片，分别为分辨率大小不同的图标资源，以便相同的应用程序在分辨率大小不同的显示窗体上都可以顺利显示。系统开始运行时会检测显示窗体的分辨率大小，自动选择与显示窗体分辨率大小匹配的目录，获取大小匹配的图标，如表1-3所示。

表1-3 4种分辨率大小不同的图标

子目录	图标分辨率大小	图例
-xhdpi	96×96	
-hdpi	72×72	
-mdpi	48×48	
-ldpi	36×36	

（2）在layout子目录中存放用户界面布局文件。该子目录中有一个系统自动生成的activity_main.xml文件，它可以按可视化的图形设计界面显示，也可以按代码设计界面显示，如图1.18所示。

activity_main.xml文件的代码如下：

```
1   <?xml version="1.0" encoding="utf-8"?>
2   <RelativeLayout xmlns:android="http://schemas.android.com/apk/res/android"
3       xmlns:tools="http://schemas.android.com/tools"
4       android:layout_width="match_parent"
5       android:layout_height="match_parent"
6       android:paddingBottom="@dimen/activity_vertical_margin"
7       android:paddingLeft="@dimen/activity_horizontal_margin"
8       android:paddingRight="@dimen/activity_horizontal_margin"
9       android:paddingTop="@dimen/activity_vertical_margin"
10      tools:context="com.example.hp480.helloandroid.MainActivity">
11      <TextView
12          android:layout_width="wrap_content"
13          android:layout_height="wrap_content"
```

```
14          android:text="Hello World!"
15          android:textSize="32sp"
16          android:id="@+id/textView" />
17  </RelativeLayout>
```

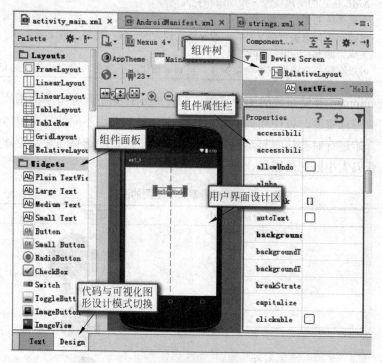

(a) 图形设计界面

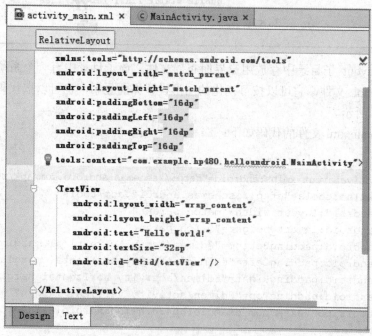

(b) 代码设计界面

图 1.18 用户界面布局文件 activity_main.xml

布局参数解析：
- "<RelativeLayout >"：相对布局配置，在相对布局中所有组件都是按前一组件的相对位置摆放。
- "android:layout_width"：定义当前视图在屏幕上所占的宽度，match_parent 即填充整个屏幕宽度。
- "android:layout_height"：定义当前视图在屏幕上所占的高度。
- "wrap_weight"：自适应大小，以便显示其全部文字内容。

在应用程序中如果使用用户界面的组件时，则需要通过 R.java 文件中的 R 类来调用。

（3）values 子目录存放参数描述文件资源。这些参数描述文件也都是 XML 文件，例如字符串（string.xml）、颜色（color.xml）、数组（arrays.xml）等。这些参数也需要通过 R.java 文件中的 R 类来调用。

3. r\debug 目录

r\debug 目录存放由系统自动产生的一个 R.java 文件，该文件将 res 目录中的资源与 id 编号进行映射，从而可以方便地对资源进行引用，如图 1.19 所示。正如该文件头部注释的说明，该文件是自动生成的，不允许用户修改。

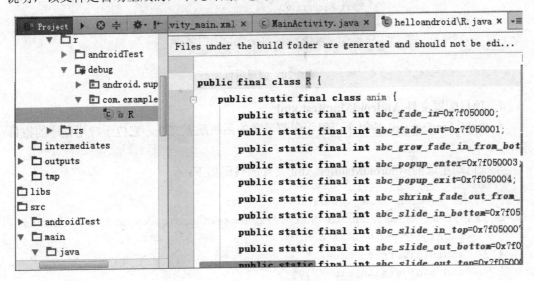

图 1.19　r\debug 目录下的 R.java 的源代码

在程序中引用资源需要使用 R 类，其引用形式如下：

R.资源文件类型.资源名称

例如：

（1）在 Activity 中显示布局视图：

```
setContentView(R.layout.main);
```

（2）程序要获得用户界面布局文件中的按钮实例 Button1：

```
mButtn = (Button)finadViewById(R.id.Button1);
```

（3）程序要获得用户界面布局文件中的文本组件实例 TextView1：

```
mEditText = (EditText)findViewById(R.id.EditText1)。
```

在编写和调试程序的过程中，有时由于操作失误会造成 MainActivity.java 程序中找不到 R 文件或显示 R 文件错误。如果出现这种情况可以按下列步骤解决：

（1）检查资源文件是否存在错误，包括 layout 文件以及图片资源等文件，如有错误及时更正。

（2）执行 Android Studio 菜单栏中的 Build→Clean Project 命令，如图 1.20 所示，经项目清理后 MainActivity.java 程序中找不到 R 文件或显示 R 文件错误的问题就能解决。

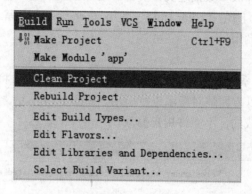

图 1.20　清理项目

4. 项目配置文件 AndroidManifest.xml

AndroidManifest.xml 是每个应用程序都需要的系统配置文件，它位于应用程序的根目录下。

系统自动生成的 AndroidManifest.xml 文件的代码如下：

```
1   <?xml version="1.0" encoding="utf-8"?>
2   <manifest xmlns:android="http://schemas.android.com/apk/res/android"
3       package="com.HelloAndroid"
4       android:versionCode="1"
5       android:versionName="1.0" >
6       <uses-sdk android:minSdkVersion="14" />
7       <application
8           android:icon="@drawable/ic_launcher"
9           android:label="@string/app_name" >
10          <activity
11              android:label="@string/app_name"
12              android:name=".MainAndroidActivity" >
13              <intent-filter >
14                  <action android:name="android.intent.action.MAIN" />
15                  <category android:name="android.intent.category.LAUNCHER"/>
16              </intent-filter>
17          </activity>
```

```
18      </application>
19  </manifest>
```

AndroidManifest.xml 文件的代码元素的说明如表 1-4 所示。

表 1-4 AndroidManifest.xml 文件的代码元素的说明

代码元素	说明
manifest	XML 文件的根结点，包含了 package 中的所有内容
xmlns:android	命名空间的声明。 xmlns:android="http://schemas.android.com/apk/res/android"使得 Android 中的各种标准属性能在文件中使用
package	声明应用程序包
uses-sdk	声明应用程序所使用的 Android SDK 版本
application	application 级别组件的根结点，声明一些全局或默认的属性，例如标签、图标、必要的权限等
android:icon	应用程序图标
android:label	应用程序名称
activity	Activity 是一个应用程序与用户交互的图形界面。每一个 Activity 必须有一个<activity>标记对应，如果一个 Activity 没有对应的标记将无法运行
android:name	应用程序默认启动的活动程序的 Activity 界面
intent-filter	声明一组组件支持的 Intent 值。在 Android 中组件之间可以相互调用，协调工作，Intent 提供组件之间通信所需要的相关信息
action	声明目标组件执行的 Intent 动作。Android 定义了一系列标准动作，例如 MAIN_ACTION、VIEW_ACTION、EDIT_ACTION 等。与此 Intent 匹配的 Activity 将会被当作进入应用的入口
category	指定目标组件支持的 Intent 类别

1.5.2 Android 应用程序结构分析

1. 逻辑控制层与表现层

从上面的 Android 应用程序可以看到，一个 Android 应用程序通常由 Activity 类程序（Java 程序）和用户界面布局 XML 文档组成。

在 Android 应用程序中逻辑控制层与表现层是分开设计的。逻辑控制层由 Java 应用程序实现，表现层由 XML 文档描述，如图 1.21 所示。

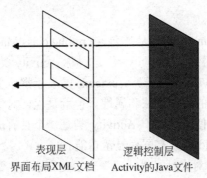

图 1.21 Android 应用程序的逻辑控制层与表现层

2. Android 程序的组成结构

Android 程序与 Java 程序的结构是相同的，打开 src 目录下的 HelloAndroid.java 文件，其代码如下：

```
1   package com.example.HelloAndroid;      ← 包声明语句

2   import android.app.Activity;
3   import android.os.Bundle;              ← 导入包
                                                        类标志
4   public class MainAndroid extends Activity  ← 类声明语句
5   {                                              类名
6       public void onCreate(Bundle savedInstanceState) ← 重写 onCreate()方法
7       {
8           super.onCreate(savedInstanceState);  ← 调用父类 Activity 的 onCreate()方法
9           setContentView(R.layout. activity_main);
10      }                                         在屏幕上显示内容的方法
11  }
```

其中：

（1）第 1 行是包声明语句，包名字是在建立应用程序的时候指定的，在这里设定为 com.example.HelloAndroid。该行的作用是指出这个文档所在的名称空间，package（包）是其关键字。使用名称空间的原因是程序一旦扩展到某个大小，程序中的变量名称、方法名称、类名等难免重复，这时就可以通过定义名称空间将定义的名称区隔，以避免相互冲突的情形发生。

（2）第 2 行和第 3 行是导入包的声明语句。这两条语句的作用是告诉系统编译器编译程序时要导入 android.app.Activity 和 android.os.Bundle 两个包。import（导入）是其关键字。在 Java 语言中使用任何 API 都要事先导入相对应的包。

（3）第 4～11 行是类的定义，这是应用程序的主体部分。Android 应用程序是由类组成的，类的一般结构如下：

```
public class MainAndroid extends Activity   //类声明
{
    … ;   //类体
}
```

class 是类的关键字，MainAndroid 是类名。在 public class MainAndroid 后面添加 extends Activity 表示 MainAndroid 类继承 Activity 类。这时称 Activity 类是 MainAndroid 类的父类，或称 MainAndroid 类是 Activity 类的子类。extends 是表示继承关系的关键字。在面向对象的程序中，子类会继承父类的所有方法和属性。也就是说，对于在父类中定义的全部方法和属性，子类可以直接拿来使用。由于 Activity 类是一个具有屏幕显示功能的活动界面程序，因此其子类 MainAndroid 也具有屏幕显示功能。

class 语句后面跟着的一对大括号"{ }"表示复合语句，是该类的主体部分，称为类体。在类体中定义类的方法和变量。

（4）第 6～10 行是在 MainAndroid 类的类体中定义一个方法。

1.6 Android 应用程序设计示例

【例 1-1】 在 AVD 模拟器中显示"我对学习 Android 很感兴趣!"。

(1)在 Android Studio 中新建一个 Android 项目,其项目名称(Application name)为 ex1_1。

(2)在系统自动生成的应用程序框架中打开修改资源目录(res\layout)中的界面布局文件 activity_main.xml,在设计界面中选中文本标签组件 TextView(组件面板中的 Ab Plain TextView 项),在属性栏中找到 text 属性,将其修改为"我对学习 Android 很感兴趣!",如图 1.22(a)所示。

(3)保存程序后,单击工具栏上的 Run App 按钮 ▶ 运行 AVD 模拟器,模拟器中的程序运行结果如图 1.22(b)所示。

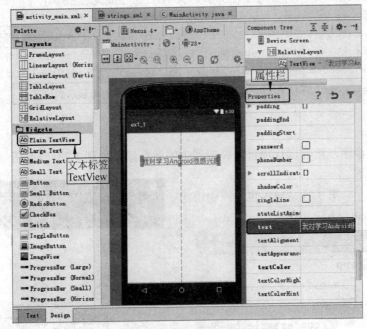

(a)修改 TextView 组件的 text 属性值

(b)在模拟器中运行的显示结果

图 1.22 在模拟器中显示"我对学习 Android 很感兴趣!"

【例 1-2】 设计一个显示资源目录中图片文件的程序。

(1)在 Android Studio 中新建一个 Android 项目,其项目名称(Application name)为 ex1_2。

(2)把事先准备的图片文件 flower.png 复制到资源目录 res\drawable 中,如图 1.23(a)所示。

(3)打开源代码目录 src 中的 MainActivity.java 文件,编写代码如下:

```
1   package com.ex1_2;
2   import android.support.v7.app.AppCompatActivity;
3   import android.os.Bundle;
4   import android.widget.ImageView;   ← 增加导入 ImageView 类的语句
5   public class MainActivity extends AppCompatActivity {
6     @Override
7     public void onCreate(Bundle savedInstanceState) {
8       super.onCreate(savedInstanceState);
9       //setContentView(R.layout.activity_main);   ← 注释该语句
10      ImageView img = new ImageView(this);   //创建ImageView对象并实例化
11      img.setImageResource(R.drawable.flower);//ImageView对象设置引用图片资源
12      setContentView(img);   ← 把 ImageView 对象显示到屏幕上
13    }
14  }
```

(4)保存程序,然后运行项目,模拟器中的运行结果如图 1.23(b)所示。

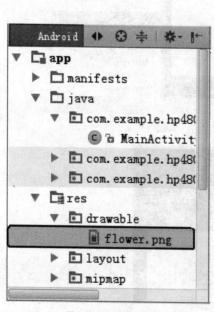

(a)导入图片后的资源目录　　　(b)程序在模拟器中运行的结果

图 1.23　在模拟器中显示图片

习 题 1

1. Android 系统是基于什么操作系统的应用系统？
2. 试述建立 Android 系统开发环境的过程和步骤。
3. 如何编写和运行 Android 系统应用程序？
4. 编写 Android 应用程序，在模拟器中显示"我对 Android 很痴迷！"。
5. 编写 Android 应用程序，在模拟器中显示一个图形文件。

第 2 章　Android 用户界面的设计

2.1　用户界面组件包 widget 和 View 类

1. 用户界面组件包 widget

Android 系统为开发人员提供了丰富多彩的用户界面组件，通过使用这些组件可以设计出炫目的界面。大多数用户界面组件放置在 widget 包中。widget 包中的常用组件如表 2-1 所示。

表 2-1　widget 包中的常用组件

可视化组件	说明
Button	按钮
CalendarView	日历视图
CheckBox	复选框
EditText	可编辑的文本输入框（对应组件面板中的"Plain Text"）
ImageView	显示图像或图标，并提供缩放、着色等图像处理方法
ListView	列表框视图
RadioGroup	单选按钮组件
Spinner	下拉列表
TextView	文本标签（对应组件面板中的"Plain TextView"）
WebView	网页浏览器视图
Toast	消息提示

2. View 类

View 类是用户界面组件的共同父类，几乎所有的用户界面组件都是继承 View 类实现的，例如 TextView、Button、EditText 等。

对于 View 类及其子类的属性，可以在布局文件 XML 中设置，也可以通过成员方法在 Java 代码文件中动态设置。View 类常用的属性和方法如表 2-2 所示。

表 2-2　View 类常用的属性和方法

属性	对应方法	说明
android:background	setBackgroundColor(int color)	设置背景颜色
android:id	setId(int)	为组件设置可通过 findViewById 方法获取的标识符
android:alpha	setAlpha(float)	设置透明度，取值范围为 0～1
	findViewById(int id)	与 id 所对应的组件建立关联
android:visibility	setVisibility(int)	设置组件的可见性
android:clickable	setClickable(boolean)	设置组件是否响应单击事件

2.2 文本标签 TextView 与按钮 Button

2.2.1 文本标签

文本标签（TextView）用于显示文本内容，是最常用的组件之一。其常用的方法见表 2-3。

表 2-3 文本标签（TextView）常用的方法

方法	功能
getText();	获取文本标签的文本内容
setText(CharSequence text);	设置文本标签的文本内容
setTextSize(float);	设置文本标签的文本大小
setTextColor(int color);	设置文本标签的文本颜色

其常用的 XML 文件元素属性见表 2-4。

表 2-4 文本标签（TextView）常用的 XML 文件元素属性

元素属性	说明
android:id	文本标签标识
android:layout_width	文本标签 TextView 的宽度，通常取值"fill_parent"（屏幕宽度）或以 pt 为单位的固定值
android:layout_height	文本标签 TextView 的高度，通常取值" wrap_content "（文本的高）或以 px 为单位的固定值
android:text	文本标签 TextView 的文本内容
android:textSize	文本标签 TextView 的文本大小

【例 2-1】 设计一个文本标签组件程序。

创建名称为 ex2_1 的新项目，包名为 com.ex2_1。打开系统自动生成的项目框架，需要设计的文件如下：

- 设计资源文件 strings.xml；
- 设计布局文件 activity_main.xml；
- 设计控制文件 MainActivity.java。

（1）设计资源文件 strings.xml。打开 res\values 下的 strings.xml，添加属性为"hello"的元素项的文本内容：

```
1 <?xml version="1.0" encoding="utf-8"?>
2 <resources>
3     <string name="hello">
4         \n  荷塘月色
5         \n  剪一段时光缓缓流淌，
6         \n  弹一首小荷淡淡的香，
7         \n  美丽的琴音就落在我身旁。
8     </string>
```

添加 hello 元素项

```
9    <string name="app_name">ex2_1</string>
10 </resources>
```

设计资源文件 strings.xml 的过程如图 2.1 所示。

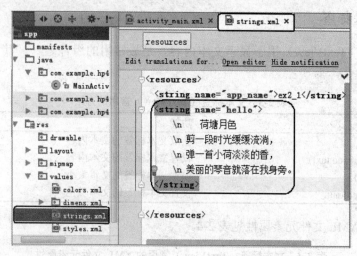

图 2.1　添加属性为"hello"的元素项的文本内容

（2）设计界面布局文件 activity_main.xml。在界面布局文件 activity_main.xml 中加入文本标签组件 TextView，设置文本标签组件的 id 属性，其 text 属性值为资源文件 strings.xml 中的 hello 项"@string/hello"，如图 2.2 所示。

完整的 activity_main.xml 代码如下：

```
1  <?xml version="1.0" encoding="utf-8"?>
2  <LinearLayout xmlns:android="http://schemas.android.com/apk/res/android"
3      android:layout_width="fill_parent"
4      android:layout_height="fill_parent"
5      android:orientation="vertical" >
6      <TextView
7          android:id="@+id/textView1"            ← 设置文本标签的 id 属性值
8          android:layout_width="fill_parent"
9          android:layout_height="wrap_content"
10         android:text="@string/hello" />
11 </LinearLayout>
```

（3）设计控制文件 MainActivity.java。在控制文件 MainActivity.java 中添加文本标签组件，并将布局文件中所定义的文本标签元素属性值赋给文本标签，与布局文件中的文本标签建立关联。源程序如下：

```
1 package com.ex2_1;
2 import android.app.Activity;
3 import android.os.Bundle;
4 import android.graphics.Color;              //引用图形颜色组件
```

```
5  import android.widget.TextView;        //引用文本标签组件
6
7  public class MainActivity extends Activity
8  {
9      private TextView txt;              ← 声明文本标签对象
10     public void onCreate(Bundle savedInstanceState)
11     {
12         super.onCreate(savedInstanceState);
13         setContentView(R.layout.activity_main);
14         txt = (TextView)findViewById(R.id.textView1);  ← 与布局文件文本标签建立关联
15         txt.setTextColor(Color.WHITE);  ← 设置文本颜色
16     }
17 }
```

保存项目，配置应用程序的运行参数，程序运行结果如图 2.3 所示。

图 2.2　在界面布局中设置文本标签

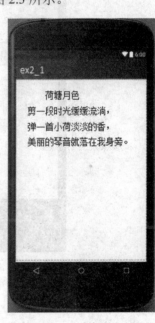

图 2.3　文本标签

2.2.2　按钮及按钮处理事件

按钮（Button）用于处理人机交互事件，在一般应用程序中经常会用到。由于按钮（Button）是文本标签 TextView 的子类，其继承关系如图 2.4 所示。按钮（Button）继承了文本标签 TextView 的所有方法和属性。

```
java.lang.Object
  └ android.view.View
      └ android.widget.TextView
          └ android.widget.Button
```

图 2.4　按钮 Button 与文本标签 TextView 的继承关系

Android 用户界面的设计

按钮（Button）在程序设计中最常用的方式是实现 OnClickListener 监听接口，当单击按钮时通过 OnClickListener 监听接口触发 onClick()事件，实现用户需要的功能。OnClickListener 接口有一个 onClick()方法，在按钮（Button）实现 OnClickListener 接口时一定要重写这个方法。

按钮（Button）调用 OnClickListener 接口对象的方法如下：

按钮对象.setOnClickListener(OnClickListener对象);

【例 2-2】 编写程序，实现在点击按钮后页面标题及文本组件的文字内容发生变化，如图 2.5 所示。

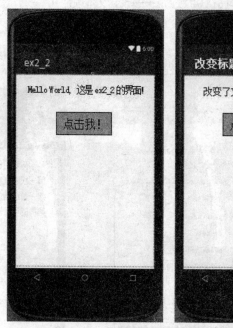

（点击按钮前）　　　　　　　　（点击按钮后）

图 2.5　点击按钮后，文本组件的文字内容发生变化

创建名称为 ex2_2 的新项目，包名为 com.ex2_2。

（1）设计布局文件 activity_main.xml。在布局文件中添加一个按钮，其 id 为 button1。

```
1  <?xml version="1.0" encoding="utf-8"?>
2  <LinearLayout xmlns:android="http://schemas.android.com/apk/res/android"
3      android:layout_width="fill_parent"
4      android:layout_height="fill_parent"
5      android:orientation="vertical" >
6      <TextView
7          android:id="@+id/textView1"         ← 设置文本标签的 id 属性值
8          android:layout_width="fill_parent"
9          android:layout_height="wrap_content"
10         android:text="@string/hello" />
```

```
11    <Button
12        android:id="@+id/button1"              ← 设置按钮的 id 属性值
13        android:layout_width="fill_parent"
14        android:layout_height="wrap_content"
15        android:text=" @string/button" />
16  </LinearLayout>
```

（2）设计控制文件 MainActivity.java。在控制文件 MainActivity.java 中设计一个实现按钮监听接口的内部类 mClick，当点击按钮时触发 onClick()事件。

```
1  package com.ex2_2;
2  import android.app.Activity;
3  import android.os.Bundle;
4  import android.view.View;
5  import android.view.View.OnClickListener;
6  import android.widget.TextView;
7  import android.widget.Button;
8
9  public class MainActivity extends Activity
10 {
11    private TextView txt;
12    private Button btn;
13    public void onCreate(Bundle savedInstanceState)
14    {
15        super.onCreate(savedInstanceState);
16        setContentView(R.layout.activity_main);
17        txt = (TextView) findViewById(R.id.textView1);    ← 与布局文件的相
18        btn = (Button)findViewById(R.id.button1);            关组件建立关联
19        btn.setOnClickListener(new mClick());   ← 注册监听接口
20    }
21      class mClick implements OnClickListener   ← 定义实现监听接口的内部类
22      {
23          public void onClick(View v)
24          {
25              MainActivity.this.setTitle("改变标题");
26              txt.setText(R.string.newStr);
27          }
28      }
29 }
```

（3）设计资源文件 strings.xml。

```
1 <?xml version="1.0" encoding="utf-8"?>
2 <resources>
3    <string name="hello">Hello World, 这是ex2_2的界面!</string>
4    <string name="app_name">ex2_2</string>
```

```
5    <string name="button">点击我!</string>
6    <string name="newStr">改变了文本标签的内容</string>
7  </resources>
```

【例2-3】 编写程序,实现点击按钮改变文本组件的文字及背景颜色,如图2.6所示。

（点击按钮前）　　　　　　　　（点击按钮后）

图2.6 点击按钮后文本组件的文字颜色发生变化

本例题涉及颜色定义,Android 系统在 android.graphics.Color 中定义了12种常见的颜色常数,见表2-5。

表2-5 常见的颜色常数

颜色常数	十六进制数色码	意义
Color.BLACK	0xff000000	黑色
Color.BLUE	0xff00ff00	蓝色
Color.CYAN	0xff00ffff	青绿色
Color.DKGRAY	0xff444444	灰黑色
Color.GRAY	0xff888888	灰色
Color.GREEN	0xff0000ff	绿色
Color.LTGRAY	0xffcccccc	浅灰色
Color.MAGENTA	0xffff00ff	红紫色
Color.RED	0xffff0000	红色
Color.TRANSPARENT	0x00ffffff	透明
Color.WHITE	0xffffffff	白色
Color.YELLOW	0xffffff00	黄色

创建名称为 ex2_3 的新项目，包名为 com.ex2_3。
(1) 设计布局文件 activity_main.xml。
在 XML 文件中表示颜色的方法有多种。
- #RGB：用 3 位十六进制数分别表示红、绿、蓝颜色。
- #ARGB：用 4 位十六进制数分别表示透明度以及红、绿、蓝颜色。
- #RRGGBB：用 6 位十六进制数分别表示红、绿、蓝颜色。
- #AARRGGBB：用 8 位十六进制数分别表示透明度以及红、绿、蓝颜色。

下面的源程序采用 8 位十六进制数表示透明度以及红、绿、蓝颜色。

```xml
1  <?xml version="1.0" encoding="utf-8"?>
2  <LinearLayout xmlns:android="http://schemas.android.com/apk/res/android"
3      android:layout_width="fill_parent"
4      android:layout_height="fill_parent"
5      android:background="#ff7f7c"
6      android:orientation="vertical" >
7      <TextView
8          android:id="@+id/textView1"
9          android:layout_width="fill_parent"
10         android:layout_height="wrap_content"
11         android:textColor="#ff000000"         ← 采用 8 位十六进制数表示颜色
12         android:text="@string/hello" />
13     <Button
14         android:id="@+id/button1"
15         android:layout_width="wrap_content"
16         android:layout_height="wrap_content"
17         android:text="@string/button" />
18 </LinearLayout>
```

(2) 设计控制文件 MainActivity.java。

```java
1  package com.ex2_3;
2  import android.app.Activity;
3  import android.graphics.Color;
4  import android.os.Bundle;
5  import android.view.View;
6  import android.view.View.OnClickListener;
7  import android.widget.Button;
8  import android.widget.TextView;
9
10 public class MainActivity extends Activity
11 {
12     /** Called when the activity is first created. */
13     private TextView txt;
```

```java
14    private Button btn;
15   @Override
16    public void onCreate(Bundle savedInstanceState)
17    {
18        super.onCreate(savedInstanceState);
19        setContentView(R.layout.activity_main);
20        btn=(Button)findViewById(R.id.button1);         // 与用户界面程序中的组件建立关联
21        txt=(TextView)findViewById(R.id.textView1);
22        btn.setOnClickListener(new click());            // 注册监听接口
23    }
24    class click implements OnClickListener              // 定义实现监听接口的内部类
25    {
26      public void onClick(View v)
27      {
28         int BLACK = 0xffcccccc;
29         txt.setText("改变了文字及背景颜色");
30         txt.setTextColor(Color.YELLOW);                // 采用颜色常数设置文字颜色
31         txt.setBackgroundColor(BLACK);                 // 设置文本标签背景颜色
32      }
33    }
34 }
```

（3）设计资源文件 strings.xml。

```xml
1 <?xml version="1.0" encoding="utf-8"?>
2 <resources>
3    <string name="hello">Hello World, MainActivity!</string>
4    <string name="app_name">Ex2_3</string>
5    <string name="button">点击我，改变文字背景颜色</string>
6 </resources>
```

2.3 文本编辑框

文本编辑框（EditText）用于接收用户输入的文本信息内容。文本编辑框 EditText 继承于文本标签 TextView，其继承关系如图 2.7 所示。

```
android.view.View
   └─ android.widget.TextView
         └─ android.widget.EditText
```

图 2.7 文本编辑框 EditText 的继承关系

文本编辑框 EditText 主要继承文本标签 TextView 的方法，其常用方法见表 2-6。

表 2-6 文本编辑框 EditText 的常用方法

方法	功能
EditText(Context context)	构造方法,创建文本编辑框对象
getText()	获取文本编辑框的文本内容
setText(CharSequence text)	设置文本编辑框的文本内容

其常用的 XML 文件元素属性见表 2-7。

表 2-7 文本编辑框 EditText 常用的 XML 文件元素

元素属性	说明
android:editable	设置是否可编辑,其值为"true"或"false"
android:numeric	设置 TextView 只能输入数字,其参数默认值为假
android:password	设置密码输入,字符显示为圆点,其值为"true"或"false"
android:phoneNumber	设置只能输入电话号码,其值为"true"或"false"

定义 EditText 的 android:numeric 属性,其取值只能是下列常量(可由"|"连接多个常量)。

- integer:可以输入数值。
- signed:可以输入带符号的数值。
- decimal:可以输入带小数点的数值。

【例 2-4】 设计一个密码验证程序,如图 2.8 所示。

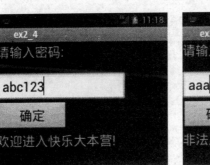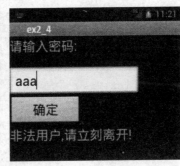

图 2.8 文本编辑框

创建名称为 ex2_4 的新项目,包名为 com.ex2_4。

(1) 设计布局文件 activity_main.xml。在界面布局中,设置一个编辑框(选择组件面板中的 Plain Text 选项),用于输入密码;再设置一个按钮,判断密码是否正确;设置两个文本标签,其中一个显示提示信息"请输入密码",另一个用于显示密码正确与否。

```
1 <?xml version="1.0" encoding="utf-8"?>
2 <LinearLayout xmlns:android="http://schemas.android.com/apk/res/android"
3     android:layout_width="fill_parent"
4     android:layout_height="fill_parent"
5     android:orientation="vertical" >
    <!--建立一个TextView -->
```

```
 6    <TextView
 7      android:id="@+id/myTextView01"
 8      android:layout_width="fill_parent"
 9      android:layout_height="41px"
10      android:layout_x="33px"
11      android:layout_y="106px"
12      android:text="请输入密码:"
13      android:textSize="24sp"
14    />
      <!--建立一个EditText -->
15    <EditText
16      android:id="@+id/myEditText"
17      android:layout_width="180px"
18      android:layout_height="wrap_content"
19      android:layout_x="29px"
20      android:layout_y="33px"
21      android:inputType="text"
22      android:textSize="24sp" />
      <!--建立一个Button -->
23    <Button
24      android:id="@+id/myButton"
25      android:layout_width="100px"
26      android:layout_height="wrap_content"
27      android:text="确定"
28      android:textSize="24sp"
29    />
      <!--建立一个TextView -->
30    <TextView
31      android:id="@+id/myTextView02"
32      android:layout_width="180px"
33      android:layout_height="41px"
34      android:layout_x="33px"
35      android:layout_y="106px"
36      android:textSize="24sp"
37    />
38  </LinearLayout>
```

（2）设计控制文件 MainActivity.java。在控制文件 MainActivity.java 中主要设计按钮的监听事件，当点击按钮后从文本编辑框中获取输入的文本内容，与密码"abc123"进行比较。

```
1 package com.ex2_4;
2 import android.app.Activity;
3 import android.os.Bundle;
4 import android.view.View;
```

```
5  import android.view.View.OnClickListener;
6  import android.widget.EditText;
7  import android.widget.TextView;
8  import android.widget.Button;

9  public class MainActivity extends Activity
10 {
11     private EditText edit;
12     private TextView txt1,txt2;
13     private Button  mButton01;
14     @Override
15     public void onCreate(Bundle savedInstanceState)
16     {
17         super.onCreate(savedInstanceState);
18         setContentView(R.layout.activity_main);
19         txt1 = (TextView)findViewById(R.id.myTextView01);    ← 与用户界面
20         txt2 = (TextView)findViewById(R.id.myTextView02);      程序中的组
21         edit = (EditText)findViewById(R.id.myEditText);        件建立关联
22         mButton01 = (Button)findViewById(R.id.myButton);
23         mButton01.setOnClickListener(new mClick());
24     }

25     class mClick implements OnClickListener    ← 定义实现监听接口的内部类
26     {
27         public void onClick(View v)
28         {
29             String passwd;
30             passwd=edit.getText().toString();    ← 获取文本编辑框中的文本内容

31             if(passwd.equals("abc123"))    ← 用equals()方法比较两个字符串是否相等
32                 txt2.setText("欢迎进入快乐大本营!");
33             else
34                 txt2.setText("非法用户,请立刻离开!");
35         }
36     }
37 }
```

2.4　Android 布局管理

　　Android 系统按照 MVC（Model-View-Controller）设计模式将应用程序的界面设计与功能控制设计分离，从而可以单独地修改用户界面，不需要修改程序代码。应用程序的用户界面通过 XML 定义组件布局来实现。

　　Android 系统的布局管理指的是在 XML 布局文件中设置组件的大小、间距、排列及对齐方式等。Android 系统中常见的布局方式有 5 种，分别是 LinearLayout、FrameLayout、TableLayout、RelativeLayout、GridLayout。组件面板中的布局组件如图 2.9 所示。

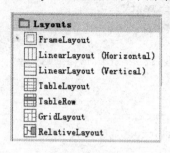

图 2.9　布局组件

2.4.1　布局文件的规范与重要属性

1. 布局文件的规范

Android 系统应用程序的 XML 布局文件有以下规范：

（1）布局文件作为应用项目的资源存放在 res\layout 目录下，其扩展名为.xml。

（2）布局文件的根结点通常是一个布局方式，在根结点内可以添加组件作为结点。

（3）布局文件的根结点必须包含一个命名空间：

xmlns:android="http://schemas.android.com/apk/res/android"

（4）如果要在实现控制功能的 Java 程序中控制界面中的组件，则必须为界面文件中的组件定义一个 id，其定义格式如下：

android:id="@+id/<组件id>"

2. 布局文件的重要属性值

在一个界面布局中会有很多元素，这些元素的大小和位置由其属性决定。下面简要介绍布局文件中的几个重要属性值。

（1）设置组件大小的属性值。

- wrap_content：根据组件内容的大小来决定组件的大小。
- match_parent：使组件填充所在容器的所有空间。

（2）设置组件大小的单位。

- px（pixels）：像素，即屏幕上的发光点。
- dp（或 dip，全称为 device independent pixels）：设备独立像素，一种支持多分辨率设备的抽象单位，和硬件相关。
- sp（scaled pixels）：比例像素，设置字体大小。

（3）设置组件的对齐方式。

在布局文件中由"android:gravity" 属性控制组件的对齐方式，其属性值有上（top）、下（bottom）、左（left）、右（right）、水平方向居中（center_horizontal）、垂直方向居中（center_vertical）等。

2.4.2 常见的布局方式

1. 线性布局

线性布局（LinearLayout）是 Android 系统中常用的布局方式之一，它将组件按照水平或垂直方向排列。在 XML 布局文件中由根元素 LinearLayout 来标识线性布局。

在布局文件中由"android:orientation"属性来控制排列方向，其属性值有水平（horizontal）和垂直（vertical）两种。

- 设置线性布局为水平方向：

```
android:orientation="horizontal"
```

- 设置线性布局为垂直方向：

```
android:orientation="vertical"
```

【例 2-5】 线性布局应用示例。

创建名称为 ex2_5 的新项目，包名为 com.ex2_5。生成项目框架后修改布局文件 activity_main.xml 如下：

```
1   <?xml version="1.0" encoding="utf-8"?>
2   <LinearLayout xmlns:android="http://schemas.android.com/apk/res/android"
3     android:layout_width="fill_parent"
4     android:layout_height="fill_parent"
5     android:orientation="vertical" >
6     <!-- android:orientation="horizontal" -->
7     <Button
8       android:id="@+id/mButton1"
9       android:layout_width="60px"
10      android:layout_height="wrap_content"
11      android:text="按钮1" />
12    <Button
13      android:id="@+id/mButton2"
14      android:layout_width="60px"
15      android:layout_height="wrap_content"
16      android:text="按钮2" />
17    <Button
18      android:id="@+id/mButton3"
19      android:layout_width="60px"
20      android:layout_height="wrap_content"
21      android:text="按钮3" />
22    <Button
23      android:id="@+id/mButton4"
24      android:layout_width="60px"
25      android:layout_height="wrap_content"
```

```
26        android:text="按钮4" />
27 </LinearLayout>
```

程序运行结果如图2.10（a）所示。如果将代码中第5行的android:orientation="vertical"（垂直方向的线性布局）更改为android:orientation="horizontal"（水平方向的线性布局），则运行结果如图2.10（b）所示。

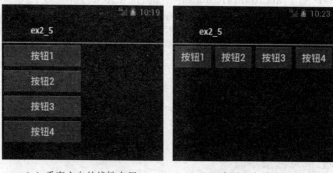

（a）垂直方向的线性布局　　（b）水平方向的线性布局

图2.10　线性布局示例

2. 帧布局

帧布局（FrameLayout）是将组件放置到左上角位置，当添加多个组件时后面的组件将遮盖之前的组件。在XML布局文件中由根元素FrameLayout来标识帧布局。

【例2-6】 帧布局应用示例。

创建名称为ex2_6的新项目，包名为com.ex2_6。生成项目框架后将事先准备的图像文件img.png复制到res\drawable目录下。

（1）设计布局文件activity_main.xml。

```
1  <?xml version="1.0" encoding="utf-8"?>
2  <FrameLayout
3    xmlns:android="http://schemas.android.com/apk/res/android"
4    android:layout_width="fill_parent"
5    android:layout_height="fill_parent">
6    <ImageView
7        android:id="@+id/mImageView"
8        android:layout_width="60px"
9        android:layout_height="wrap_content"
10   />
11   <TextView
12       android:layout_width="wrap_content"
13       android:layout_height="wrap_content"
14       android:text="快乐大本营"
15       android:textSize="18sp"
16   />
17 </FrameLayout>
```

（2）设计控制文件 MainActivity.java。

```
1   package com.ex2_6;
2   import android.app.Activity;
3   import android.os.Bundle;
4   import android.widget.ImageView;
5   public class MainActivity extends Activity
6   {
7       ImageView imageview;
8       @Override
9       public void onCreate(Bundle savedInstanceState)
10      {
11          super.onCreate(savedInstanceState);
12          setContentView(R.layout.activity_main);
13          imageview = (ImageView) this.findViewById(R.id.mImageView);
14          imageview.setImageResource(R.drawable.img);
15      }
16  }
```

程序运行结果如图 2.11 所示，可见布局文件中后添加的文本框组件遮挡了之前的图像组件。

3. 表格布局

表格布局（TableLayout）是将页面划分成由行、列构成的单元格。在 XML 布局文件中由根元素 TableLayout 来标识表格布局。

表格的列数由 android:shrinkColumns 定义，例如 android:shrinkColumns = "0, 1, 2"，即表格为 3 列，其列编号为第 1、2、3 列。

表格的行由 <TableRow> </TableRow> 定义。组件放置到哪一列由 android:layout_column 指定列编号。

【例 2-7】表格布局应用示例。设计一个 3 行 4 列的表格布局，组件安排如图 2.12 所示。

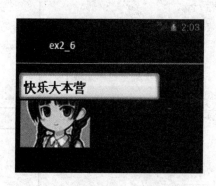

图 2.11　帧布局示例

图 2.12　3 行 4 列的表格布局

创建名称为 ex2_7 的新项目，包名为 com.ex2_7。生成项目框架后将准备好的图像文件 img1.png～img5.png 复制到 res\drawable-hdpi 目录下。

（1）设计表格布局的布局文件 activity_main.xml。在图 2.12 所示的界面布局中有用来显示图片的空白单元格，这时可以使用文本标签组件将其文字内容设置为空，这样显示出来的就是空白单元格的形式。

```xml
1  <?xml version="1.0" encoding="utf-8"?>
2  <TableLayout xmlns:android="http://schemas.android.com/apk/res/android"
3      android:layout_width="fill_parent"
4      android:layout_height="fill_parent">
5  <TableRow>     <!-- 第1行 -->
6  <ImageView  android:id="@+id/mImageView1"       ← 第1列
7      android:layout_width="wrap_content "
8      android:layout_height="wrap_content"
9      android:src="@drawable/img1" />
10 <ImageView  android:id="@+id/mImageView2"      ← 第2列          ← 第1行
11     android:layout_width="wrap_content "
12     android:layout_height="wrap_content"
13     android:src="@drawable/img2" />
14 </TableRow>

15 <TableRow>     <!-- 第2行 -->
16 <TextView
17     android:id="@+id/textView1"              ← 第1列空白单元格
18     android:layout_width="wrap_content"
19     android:layout_height="wrap_content" />
20 <ImageView  android:id="@+id/mImageView3"     ← 第2列
21     android:layout_width="wrap_content "
22     android:layout_height="wrap_content"
23     android:src="@drawable/img3" />
24 <ImageView  android:id="@+id/mImageView4"     ← 第3列            ← 第2行
25     android:layout_width=" wrap_content "
26     android:layout_height="wrap_content"
27     android:src="@drawable/img4" />
28 </TableRow>
29 <TableRow>     <!-- 第3行 -->
30 <TextView
31     android:id="@+id/textView2"              ← 第1列空白单元格
32     android:layout_width="wrap_content"
33     android:layout_height="wrap_content" />
34 <TextView                                                      ← 第3行
35     android:id="@+id/textView3"              ← 第2列空白单元格
36     android:layout_width="wrap_content"
37     android:layout_height="wrap_content" />
38 <TextView
```

```
39        android:id="@+id/textView4"              ← 第3列空白单元格
40        android:layout_width="wrap_content"
41        android:layout_height="wrap_content" />
42      <ImageView  android:id="@+id/mImageView5"  ← 第4列
43        android:layout_width="wrap_content"                    ← 第3行
44        android:layout_height="wrap_content"
45        android:src="@drawable/img5" />
46    </TableRow>
47 </TableLayout>
```

（2）设计控制文件 MainActivity.java。

```
1  package com.ex2_7;
2  import android.app.Activity;
3  import android.os.Bundle;
4  import android.widget.ImageView;

5  public class MainActivity extends Activity
6  {
7      ImageView img1, img2, img3, img4, img5;
8      @Override
9      public void onCreate(Bundle savedInstanceState)
10     {
11         super.onCreate(savedInstanceState);
12         setContentView(R.layout.activity_main);
13         img1 = (ImageView) this.findViewById(R.id.mImageView1);
14         img2 = (ImageView) this.findViewById(R.id.mImageView2);
15         img3 = (ImageView) this.findViewById(R.id.mImageView3);
16         img4 = (ImageView) this.findViewById(R.id.mImageView4);
17         img5 = (ImageView) this.findViewById(R.id.mImageView5);
18         img1.setImageResource(R.drawable.img1);
19         img2.setImageResource(R.drawable.img2);
20         img3.setImageResource(R.drawable.img3);
21         img4.setImageResource(R.drawable.img4);
22         img5.setImageResource(R.drawable.img5);
23     }
24 }
```

4. 相对布局

相对布局（RelativeLayout）是采用相对其他组件的位置的布局方式。在相对布局中通过指定 id 关联其他组件与之右对齐、上下对齐或以屏幕中央等方式来排列组件。

在 XML 布局文件中由根元素 RelativeLayout 来标识相对布局。由于相对布局属性比较多，下面简单介绍几个常用属性。

（1）设置距父元素右对齐：

```
android:layout_alignParentRight="true"
```

（2）设置该控件在 id 为 re_edit_0 的控件的下方：

```
android:layout_below="@id/re_edit_0"
```

（3）设置该控件在 id 为 re_image_0 的控件的左边：

```
android:layout_toLeftOf="@id/re_iamge_0"
```

（4）设置当前控件与 id 为 name 的控件的上方对齐：

```
android:layout_alignTop="@id/name"
```

（5）设置偏移的像素值：

```
android:layout_marginRight="30dip"
```

下面简单归纳一下。
- 第 1 类：属性值为 true 或 false。

```
android:layout_centerHrizontal
android:layout_centerVertical
android:layout_centerInparent
android:layout_alignParentBottom
android:layout_alignParentLeft
android:layout_alignParentRight
android:layout_alignParentTop
android:layout_alignWithParentIfMissing
```

- 第 2 类：属性值为 id 的引用名 "@id/id-name"。

```
android:layout_below
android:layout_above
android:layout_toLeftOf
android:layout_toRightOf
android:layout_alignTop
```

- 第 3 类：属性值为具体的像素值，例如 30dip。

```
android:layout_marginBottom
android:layout_marginLeft
android:layout_marginRight
android:layout_marginTop
```

【例 2-8】 应用相对布局设计一个组件排列如图 2.13 所示的应用程序。

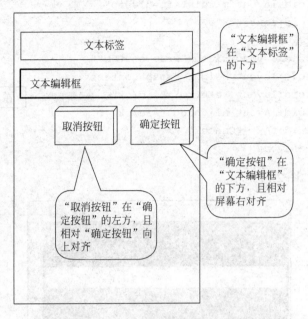

图 2.13　应用相对布局设计组件排列

创建名称为 ex2_8 的新项目，包名为 com.ex2_8。生成项目框架后修改布局文件 activity_main.xml 如下：

```
1   <?xml version="1.0" encoding="utf-8"?>
2   <RelativeLayout xmlns:android="http://schemas.android.com/apk/res/android"
3       android:layout_width="fill_parent"
4       android:layout_height="fill_parent">
5       <TextView
6           android:id="@+id/label"
7           android:layout_width="fill_parent"
8           android:layout_height="wrap_content"
9           android:textSize="24sp"
10          android:text="相对布局"/>
11      <EditText
12          android:id="@+id/edit"
13          android:layout_width="fill_parent"
14          android:layout_height="wrap_content"
15          android:background="@android:drawable/editbox_background"
16          android:layout_below="@id/label"/>      ← 在文本标签的下方
17      <Button
18          android:id="@+id/ok"
19          android:layout_width="wrap_content"
20          android:layout_height="wrap_content"
21          android:layout_below="@id/edit"         ← 在文本编辑框的下方
```

```
22        android:layout_alignParentRight="true"
23        android:layout_marginLeft="10dip"
24        android:text="OK" />
25    <Button
26        android:layout_width="wrap_content"
27        android:layout_height="wrap_content"
28        android:layout_toLeftOf="@id/ok"        ← 在 OK 按钮的左方
29        android:layout_alignTop="@id/ok"
30        android:text="Cancel" />
31  </RelativeLayout>
```

程序运行结果如图 2.14 所示。

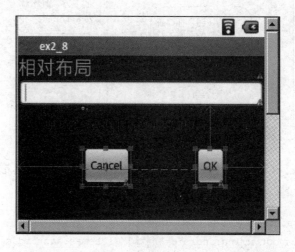

图 2.14　相对布局

5. 网格布局

网格布局（GridLayout）是把设置区域划分为若干行和若干列的网格，与 TableLayout 类似，在网格的内部放置各种需要的组件。

应用网格布局的属性可以设置组件在网格中的大小和摆放方式，网格中的一个组件可以占据多行或多列。

网格布局的主要属性如下。

- alignmentMode：设置布局管理器的对齐方式。
- columnCount：设置网格列的数量。
- rowCount：设置网格行的数量。
- layout_columnSpan：设置组件占据的列数。
- layout_rowSpan：设置组件占据的行数。

【例 2-9】 应用网格布局设计一个计算器界面。

计算器的设计界面如图 2.15 所示。在界面设计区域中设置一个 6 行 4 列的网格布局，第 1 行为显示数据的文本标签，第 2 行为清除数据的按钮，

第3~6行均划分为4列，共安排16个按钮，分别代表数字0、1、2、……、9及加、减、乘、除、等号等符号。

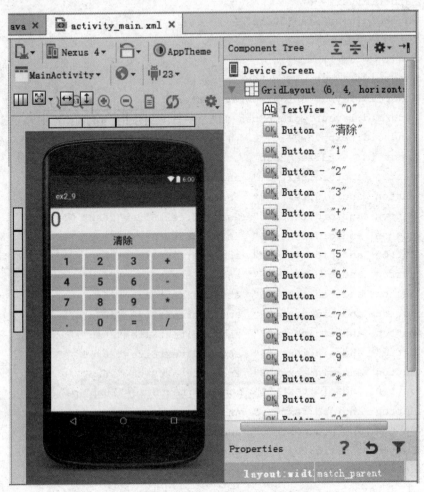

图2.15 应用网格布局设计计算器的界面

修改布局文件activity_main.xml如下：

```
1    <?xml version="1.0" encoding="utf-8"?>
2    <GridLayout xmlns:android="http://schemas.android.com/apk/res/android"
3      android:layout_width="match_parent"
4      android:layout_height="match_parent"
5      android:rowCount="6"              ← 设置网格为6行4列
6      android:columnCount="4"
7      >
8      <!--文本标签-->
9      <TextView
10       android:layout_width="wrap_content"
11       android:layout_height="wrap_content"
12       android:layout_columnSpan="4"    ← 该组件占据4列的位置
```

```xml
13      android:layout_marginLeft="4px"
14      android:gravity="left"
15      android:text="0"
16      android:textSize="50dip"
17   />
18   <Button
19      android:layout_width="match_parent"
20      android:layout_height="wrap_content"
21      android:layout_columnSpan="4"    ← 该组件占据 4 列的位置
22      android:text="清除"
23      android:textSize="26sp" />
24   <Button  android:text="1"  android:textSize="26sp" />
25   <Button  android:text="2"  android:textSize="26sp" />
26   <Button  android:text="3"  android:textSize="26sp" />
27   <Button  android:text="+"  android:textSize="26sp" />
28   <Button  android:text="4"  android:textSize="26sp" />
29   <Button  android:text="5"  android:textSize="26sp" />
30   <Button  android:text="6"  android:textSize="26sp" />
31   <Button  android:text="-"  android:textSize="26sp" />
32   <Button  android:text="7"  android:textSize="26sp" />
33   <Button  android:text="8"  android:textSize="26sp" />
34   <Button  android:text="9"  android:textSize="26sp" />
35   <Button  android:text="*"  android:textSize="26sp" />
36   <Button  android:text="."  android:textSize="26sp" />
37   <Button  android:text="0"  android:textSize="26sp" />
38   <Button  android:text="="  android:textSize="26sp" />
39   <Button  android:text="/"  android:textSize="26sp" />
40   </GridLayout>
```

2.5 进度条和选项按钮

2.5.1 进度条

进度条（ProgressBar）能以形象的图示方式直观显示某个过程的进度。进度条的常用属性和方法见表 2-8。

表 2-8　进度条的常用属性和方法

属性	方法	功能
android:max	setMax(int max)	设置进度条的变化范围为 0～max
android:progress	setProgress(int progress)	设置进度条的当前值（初始值）
	incrementProgressBy(int diff)	设置进度条的变化步长值

【例 2-10】 进度条应用示例。

在界面设计中安排一个进度条组件,并设置两个按钮,用于控制进度条的进度变化,如图 2.16 所示。

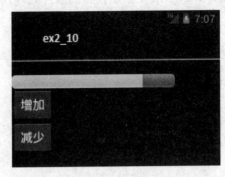

图 2.16 进度条的进度控制

程序设计步骤:

(1) 在布局文件中声明 ProgressBar。

(2) 在 Activity 中获得 ProgressBar 实例。

(3) 调用 ProgressBar 的 incrementProgressBy()方法增加和减少进度。

程序代码设计:

(1) 设计布局文件 activity_main.xml。

```xml
1  <?xml version="1.0" encoding="utf-8"?>
2  <LinearLayout xmlns:android="http://schemas.android.com/apk/res/android"
3      android:layout_width="fill_parent"
4      android:layout_height="fill_parent"
5      android:orientation="vertical" >
6      <ProgressBar
7          android:id="@+id/ProgressBar01"
8          style="@android:style/Widget.ProgressBar.Horizontal"
9      android:layout_width="250dip"
10     android:layout_height="wrap_content"
11     android:max="200"
12     android:progress="50" >
13     </ProgressBar>
14     <Button
15     android:id="@+id/button1"
16     android:layout_width="wrap_content"
17     android:layout_height="wrap_content"
18     android:text="@string/btn1" />
19     <Button
20     android:id="@+id/button2"
21     android:layout_width="wrap_content"
22     android:layout_height="wrap_content"
```

```
23    android:text="@string/btn2" />
24 </LinearLayout>
```

（2）在控制程序 MainActivity.java 中添加按钮的事件处理代码。

```
1  package com.ex2_10;
2  import android.app.Activity;
3  import android.os.Bundle;
4  import android.view.View;
5  import android.view.View.OnClickListener;
6  import android.widget.Button;
7  import android.widget.ProgressBar;

8  public class MainActivity extends Activity
9  {
10   ProgressBar progressBar;
11   Button btn1,btn2;
12    @Override
13   public void onCreate(Bundle savedInstanceState)
14   {
15     super.onCreate(savedInstanceState);
16     setContentView(R.layout.activity_main);
17     progressBar = (ProgressBar)findViewById(R.id.ProgressBar01);
18     btn1=(Button)findViewById(R.id.button1);      ← 与用户界面程序中的组件建立关联
19     btn2=(Button)findViewById(R.id.button2);
20     btn1.setOnClickListener(new click1());        ← 注册按钮的监听器
21     btn2.setOnClickListener(new click2());
22   }
23   class click1 implements OnClickListener
24   {
25     public void onClick(View v)
26       { progressBar.incrementProgressBy(5); }    ← 进度增加
27   }
28   class click2 implements OnClickListener
29   {
30     public void onClick(View v)
31       { progressBar.incrementProgressBy(-5); }   ← 进度减少
32   }
33 }
```

2.5.2 选项按钮

1. 复选按钮 CheckBox

复选按钮 CheckBox 用于多项选择的情形，用户可以一次性选择多个选项。复选按钮

CheckBox 是按钮 Button 的子类，其属性与方法继承于按钮 Button。复选按钮 CheckBox 的常用方法见表 2-9。

表 2-9 复选按钮 CheckBox 的常用方法

方法	功能
isChecked()	判断选项是否被选中
getText()	获取复选按钮的文本内容

【例 2-11】 复选按钮应用示例。

在界面设计中安排 3 个复选按钮和一个普通按钮，选择选项后点击按钮，在文本标签中显示所选中选项的文本内容。如图 2.17 所示。

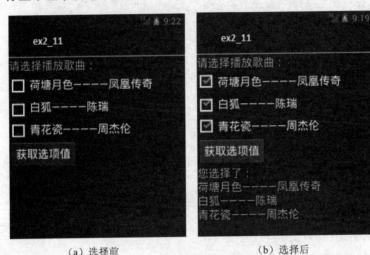

(a) 选择前　　　　　　　　(b) 选择后

图 2.17 复选按钮的多项选择

程序设计步骤：
（1）在布局文件中声明复选按钮 CheckBox。
（2）在 Activity 中获得复选按钮 CheckBox 实例。
（3）调用 CheckBox 的 isChecked()方法判断该选项是否被选中，如果被选中，则调用 getText()方法获取选项的文本内容。

程序代码设计：
（1）设计布局文件 activity_main.xml。

```
1  <?xml version="1.0" encoding="utf-8"?>
2  <LinearLayout xmlns:android="http://schemas.android.com/apk/res/android"
3      android:layout_width="fill_parent"
4      android:layout_height="fill_parent"
5      android:orientation="vertical" >
6      <TextView
7          android:layout_width="fill_parent"
8          android:layout_height="wrap_content"
```

```xml
9       android:text="@string/hello"
10      android:textSize="20sp"/>
11   <CheckBox
12       android:id="@+id/check1"
13       android:layout_width="fill_parent"
14       android:layout_height="wrap_content"
15  android:textSize="20sp"
16  android:text="@string/one" />
17   <CheckBox
18       android:id="@+id/check2"
19       android:layout_width="fill_parent"
20       android:layout_height="wrap_content"
21       android:textSize="20sp"
22       android:text="@string/two" />
23   <CheckBox
24       android:id="@+id/check3"
25       android:layout_width="fill_parent"
26       android:layout_height="wrap_content"
27       android:textSize="20sp"
28       android:text="@string/three" />
29   <Button
30       android:id="@+id/button"
31       android:layout_width="wrap_content"
32       android:layout_height="wrap_content"
33       android:textSize="20sp"
34       android:text="@string/btn" />
35   <TextView
36        android:id="@+id/textView2"
37        android:layout_width="fill_parent"
38   android:layout_height="wrap_content"
39   android:text=""
40   android:textSize="20sp"/>
41  </LinearLayout>
```

（2）在 strings.xml 文件中设置要使用的字符串。

```xml
1 <?xml version="1.0" encoding="utf-8"?>
2 <resources>
3     <string name="hello">请选择播放歌曲：</string>
4     <string name="app_name">ex2_11</string>
5     <string name="one">荷塘月色————凤凰传奇</string>
6     <string name="two">白狐————陈瑞</string>
7     <string name="three">青花瓷————周杰伦</string>
8     <string name="btn">获取选项值</string>
9 </resources>
```

(3) 设计控制程序 MainActivity.java。在控制程序 MainActivity.java 中建立程序中组件与用户界面程序组件的关联，并编写按钮的事件处理代码。

```java
1  package com.ex2_11;
2  import android.app.Activity;
3  import android.os.Bundle;
4  import android.view.View;
5  import android.view.View.OnClickListener;
6  import android.widget.Button;
7  import android.widget.CheckBox;
8  import android.widget.TextView;
9  public class MainActivity extends Activity
10 {
11    CheckBox ch1,ch2,ch3;
12    Button okBtn;
13    TextView txt;
14    @Override
15  public void onCreate(Bundle savedInstanceState)
16  {
17    super.onCreate(savedInstanceState);
18    setContentView(R.layout.activity_main);
19    ch1=(CheckBox)findViewById(R.id.check1);
20    ch2=(CheckBox)findViewById(R.id.check2);
21    ch3=(CheckBox)findViewById(R.id.check3);
22    okBtn=(Button)findViewById(R.id.button);
23    txt=(TextView)findViewById(R.id.textView2);
24    okBtn.setOnClickListener(new click());
25  }

26  class click implements OnClickListener
27  {
28     public void onClick(View v)
29     {
30  String str="";
31  if(ch1.isChecked()) str=str+"\n"+ch1.getText();
32  if(ch2.isChecked()) str=str+"\n"+ch2.getText();
33      if(ch3.isChecked()) str=str+"\n"+ch3.getText();
34  txt.setText("您选择了："+str);
35    }
36  }
37 }
```

← 与用户界面程序中的组件建立关联

2. 单选组件 RadioGroup 与单选按钮 RadioButton

单选组件 RadioGroup 用于多项选择中只允许任选其中一项的情形。单选组件

RadioGroup 由一组单选按钮 RadioButton 组成。单选按钮 RadioButton 是按钮 Button 的子类。单选按钮 RadioButton 的常用方法见表 2-10。

表 2-10 单选按钮 RadioButton 的常用方法

方法	功能
isChecked()	判断选项是否被选中
getText()	获取单选按钮的文本内容

【例 2-12】 单选按钮应用示例。

在界面设计中安排两个单选按钮、一个文本编辑框和一个普通按钮，选择选项后点击按钮，在文本标签中显示文本编辑框及所选中选项的文本内容，如图 2.18 所示。

图 2.18 单选按钮示例

程序设计步骤：

（1）在布局文件中声明单选组件 RadioGroup 和单选按钮 RadioButton。

（2）在 Activity 中获得单选按钮 RadioButton 实例。

（3）调用 RadioButton 的 isChecked()方法判断该选项是否被选中，如果被选中，则调用 getText()方法获取选项的文本内容。

程序代码设计：

（1）设计布局文件 activity_main.xml。

```
1  <?xml version="1.0" encoding="utf-8"?>
2  <LinearLayout xmlns:android="http://schemas.android.com/apk/res/android"
3      android:layout_width="fill_parent"
4      android:layout_height="fill_parent"
5      android:orientation="vertical" >
6      <TextView
7          android:layout_width="fill_parent"
8          android:layout_height="wrap_content"
```

```
9       android:textSize="20sp"
10      android:text="@string/hello" />
11  <EditText
12      android:id="@+id/edit1"
13      android:layout_width="fill_parent"
14      android:layout_height="wrap_content"
15      android:inputType="text"
16      android:textSize="20sp" />
17  <RadioGroup
18      android:layout_width="fill_parent"
19      android:layout_height="wrap_content">
20  <RadioButton
21      android:id="@+id/boy01"
22      android:text="@string/boy"/>
23  <RadioButton
24      android:id="@+id/girl01"
25      android:text="@string/girl" />
26  </RadioGroup>
27  <Button
28      android:id="@+id/myButton"
29      android:layout_width="wrap_content"
30      android:layout_height="wrap_content"
31      android:text="确定"
32      android:textSize="20sp"
33      />
34  <TextView
35      android:id="@+id/text02"
36      android:layout_width="fill_parent"
37      android:layout_height="wrap_content"
38      android:textSize="20sp"
39      />
40  </LinearLayout>
```

（2）在 strings.xml 文件中设置要使用的字符串。

```
1 <?xml version="1.0" encoding="utf-8"?>
2 <resources>
3   <string name="hello">请输入您的姓名：</string>
4   <string name="app_name">ex2_12</string>
5   <string name="boy">男</string>
6   <string name="girl">女</string>
7 </resources>
```

(3)设计控制程序 MainActivity.java。在控制程序 MainActivity.java 中建立程序中组件与用户界面程序组件的关联，并编写按钮的事件处理代码。

```java
 1  package com.ex2_12;
 2  import android.app.Activity;
 3  import android.os.Bundle;
 4  import android.view.View;
 5  import android.view.View.OnClickListener;
 6  import android.widget.Button;
 7  import android.widget.EditText;
 8  import android.widget.RadioButton;
 9  import android.widget.TextView;
10  public class MainActivity extends Activity
11  {
12    Button okBtn;
13    EditText edit;
14    TextView txt;
15    RadioButton r1, r2;
16    @Override
17    public void onCreate(Bundle savedInstanceState)
18    {
19       super.onCreate(savedInstanceState);
20       setContentView(R.layout.activity_main);
21       edit = (EditText) findViewById(R.id.edit1);
22       okBtn = (Button) findViewById(R.id.myButton);           ← 与用户界面程序中
23       txt = (TextView) findViewById(R.id.text02);               的组件建立关联
24       r1 = (RadioButton) findViewById(R.id.boy01);
25       r2 = (RadioButton) findViewById(R.id.girl01);
26       okBtn.setOnClickListener(new mClick());
27     }
28     class mClick implements OnClickListener
29     {
30        public void onClick(View v)
31        {
32          CharSequence str = "", name = "";
33          name = edit.getText();
34          if (r1.isChecked())      ← 第 1 个单选按钮被选中
35             str = r1.getText();
36          if (r2.isChecked())      ← 第 2 个单选按钮被选中
37             str = r2.getText();
38          txt.setText("您输入的信息为：\n 姓名 " + name + "\t 性别 " + str);
```

```
39      }
40    }
41 }
```

2.6 图像显示类 ImageView 与画廊组件类 Gallery

2.6.1 图像显示类 ImageView

ImageView 类用于显示图片或图标等图像资源，并提供图像缩放及着色（渲染）等图像处理功能。

ImageView 类的常用属性和对应方法见表 2-11。

表 2-11 ImageView 类的常用属性和方法

元素属性	对应方法	说明
android:maxHeight	setMaxHeight(int)	为显示图像提供最大高度的可选参数
android:maxWidth	setMaxWidth(int)	为显示图像提供最大宽度的可选参数
android:scaleType	setScaleType(ImageView.ScaleType)	控制图像适合 ImageView 大小的显示方式（参见表 2-12）
android:src	setImageResource(int)	获取图像文件的路径

ImageView 类的 scaleType 属性值见表 2-12。

表 2-12 ImageView 类的 scaleType 属性值

scaleType 属性值常量	值	说明
matrix	0	用矩阵来绘图
fitXY	1	拉伸图片（不按宽高比例）以填充 View 的宽高
fitStart	2	按比例拉伸图片，拉伸后图片的高度为 View 的高度，且显示在 View 的左边
fitCenter	3	按比例拉伸图片，拉伸后图片的高度为 View 的高度，且显示在 View 的中间
fitEnd	4	按比例拉伸图片，拉伸后图片的高度为 View 的高度，且显示在 View 的右边
center	5	按原图大小显示图片，当图片宽高大于 View 的宽高时截取图片中间部分显示
centerCrop	6	按比例放大原图直至等于某边 View 的宽高显示
centerInside	7	当原图宽高等于 View 的宽高时按原图大小居中显示，否则将原图缩放至 View 的宽高居中显示

【例 2-13】 显示图片示例。

程序设计步骤：

（1）将事先准备好的图片序列 img1.jpg～img6.jpg 复制到 res\drawable 目录下。

（2）在布局文件中声明图像显示组件 ImageView。

(3) 在 Activity 中获得相关组件实例。
(4) 通过触发按钮事件调用 OnClickListener 接口的 onClick()方法显示图像。
程序代码设计：
(1) 设计用户界面程序 activity_main.xml。在界面设计中，安排两个按钮和一个图像显示组件 ImageView，点击按钮可以翻阅浏览图片。其布局设计如图 2.19 所示。

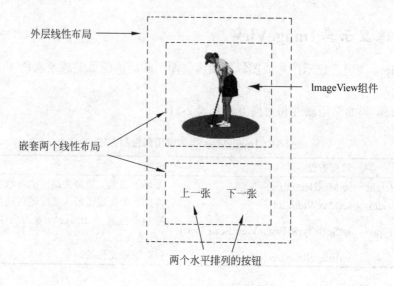

图 2.19　布局设计

安排好组件后需要设置图像显示组件的数据源，其步骤如下：
① 选取 ImageView 组件，在属性（Properties）栏中选择 src 选项，然后点击右边的按钮设置显示图像的数据源，如图 2.20 所示。

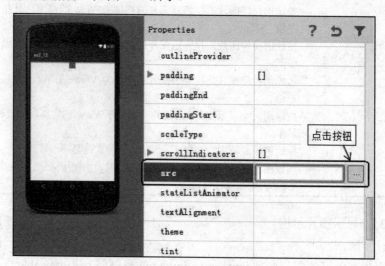

图 2.20　设置显示图像的数据源

② 在弹出的 Resources 对话框中选择 Drawable 选项，选取复制在 res\drawable 目录下

的图像，如图 2.21 所示。

图 2.21　选取复制在 res\drawable 目录下的图像

用户界面程序 activity_main.xml 的代码如下：

```
1  <?xml version="1.0" encoding="utf-8"?>
2  <LinearLayout xmlns:android="http://schemas.android.com/apk/res/android"
3   android:layout_width="fill_parent"
4   android:layout_height="fill_parent"
5   android:gravity="center|fill"
6   android:orientation="vertical" >
7   <LinearLayout
8    android:layout_width="fill_parent"
9    android:layout_height="wrap_content"
10   android:gravity="center" >
11   <ImageView
12    android:id="@+id/img"
13    android:layout_width="240dip"
14    android:layout_height="240dip"
15    android:layout_centerVertical="true"
16    android:src="@drawable/img1" />
17  </LinearLayout>
18    <LinearLayout
19    android:layout_width="fill_parent"
20    android:layout_height="wrap_content" >
21    <Button
22      android:id="@+id/btn_last"
```

```
23      android:layout_width="150dip"
24      android:layout_height="wrap_content"
25      android:text="上一张" />
26    <Button android:id="@+id/btn_next"
27      android:layout_width="150dip"
28      android:layout_height="wrap_content"
29      android:text="下一张" />
30  </LinearLayout>
31 </LinearLayout>
```

（2）设计控制程序 MainActivity.java。在控制程序 MainActivity.java 中建立程序中组件与用户界面程序组件的关联，并编写按钮的事件处理代码。

```
1  package com.ex2_13;
2  import android.app.Activity;
3  import android.os.Bundle;
4  import android.view.View;
5  import android.view.View.OnClickListener;
6  import android.widget.Button;
7  import android.widget.ImageView;
8  public class MainActivity extends Activity {
9  ImageView img;
10 Button btn_last, btn_next;
11 //存放图片id的int数组
12 private int[] imgs={
13 R.drawable.img1,
14 R.drawable.img2,
15 R.drawable.img3,          ← 数组元素为资源目录中的图片序列
16 R.drawable.img4,            img1.jpg、img2.jpg、…、img6.jpg
17 R.drawable.img5,
18 R.drawable.img6 };
19 int index=1;
20  @Override
21  public void onCreate(Bundle savedInstanceState) {
22  super.onCreate(savedInstanceState);
23  setContentView(R.layout.activity_main);
24  img = (ImageView)findViewById(R.id.img);
25  btn_last = (Button)findViewById(R.id.btn_last);   ← 与用户界面程序中
26  btn_next = (Button)findViewById(R.id.btn_next);     的组件进行关联
27  btn_last.setOnClickListener(new mClick());
28  btn_next.setOnClickListener(new mClick());        ← 注册监听接口
29   }
30   class mClick implements OnClickListener   //定义一个类实现监听接口
31   {
32    public void onClick(View v)
```

```
33    {
34     if(v==btn_last)
35     {
36      if(index>0 && index<imgs.length)
37      {
38       index--;
39       img.setImageResource(imgs[index]);
40       } else {index=imgs.length+1;  }
41     }
42      if(v==btn_next)
43      {
44       if(index>0&&index<imgs.length-1)
45       {
46        index++;
47        img.setImageResource(imgs[index]);
48        }else  {index=imgs.length-1;       }
49      }
50     }
51    }
52   }
```

"上一张"按钮事件

"下一张"按钮事件

程序的运行结果如图 2.22 所示。

图 2.22 图像显示示例

2.6.2 画廊组件类 Gallery 与图片切换器 ImageSwitcher

Gallery 类是 Android 中控制图片展示的一个组件，它可以横向显示一列图像。Gallery 类的常用属性及方法见表 2-13。

表 2-13 Gallery 类的常用属性及方法

元素属性	对应方法	说明
android:spacing	setSpacing(int)	设置图片之间的间距，以像素为单位
android:unselectedAlpha	setUnselectedAlpha(float)	设置未选中图片的透明度（Alpha）
android:animationDuration	setAnimationDuration（int）	设置布局变化时动画的转换所需的时间（毫秒级），仅在动画开始时计时
	onTouchEvent(MotionEvent event)	触摸屏幕时触发 MotionEvent 事件
	onDown(MotionEvent e)	按下屏幕时触发 MotionEvent 事件

Gallery 类经常与图片切换器 ImageSwitcher 配合使用，用图片切换器 ImageSwitcher 类展示图片效果。在使用 ImageSwitcher 时必须用 ViewFactory 接口的 makeView()方法创建视图。ImageSwitcher 的常用方法见表 2-14。

表 2-14 ImageSwitcher 的常用方法

方法	说明
setInAnimation (Animation inAnimation)	设置绘制动画对象进入屏幕的方式
setOutAnimation(Animation outAnimation)	设置绘制动画对象退出屏幕的方式
setImageResource(int resid)	设置显示的初始图片
showNext()	显示下一个视图
showPrevious()	显示前一个视图

【例 2-14】画廊展示图片示例。

在界面设计中安排一个画廊组件 Gallery 和一个图片切换器 ImageSwitcher，实现点击画廊中的小图片可以在图片切换器中显示放大的图片，如图 2.23 所示。

图 2.23 画廊展示图片示例

程序设计步骤：

（1）在布局文件中声明画廊组件 Gallery 和图片切换器 ImageSwitcher，采用表格布局。

（2）把事先准备好的图片文件 img1.jpg～img8.jpg 复制到项目的资源目录 res\drawable 中，在 Activity 中创建一个图像文件数组 imgs[]，其数组的元素为图片文件。

（3）在 Activity 中创建画廊组件 Gallery 和图片切换器 ImageSwitcher 组件的实例对象。

（4）在 Activity 中创建一个实现 ViewFactory 接口的内部类，重写 makeView()方法建立 imageView 图像视图。图片切换器 ImageSwitcher 通过该图像视图显示放大的图片。

（5）在 Activity 中创建一个 BaseAdapter 适配器子类的内部类，用于安排放在画廊 gallery 中的图片文件及显示方式。

程序代码设计：

（1）设计用户界面程序 activity_main.xml。

```xml
1  <?xml version="1.0" encoding="utf-8"?>
2  <TableLayout android:id="@+id/TableLayout01"
3    android:layout_width="wrap_content"
4    android:layout_height="wrap_content"
5    xmlns:android="http://schemas.android.com/apk/res/android"
6    android:layout_gravity="center">
7  <TextView
8    android:layout_width="fill_parent"
9    android:layout_height="wrap_content"
10   android:textSize="20sp"
11   android:text="@string/hello" />
12 <Gallery android:id="@+id/Gallery01"
13   android:layout_width="wrap_content"
14   android:layout_height="wrap_content"
15   android:spacing="10dip"/>
16 <ImageSwitcher android:id="@+id/ImageSwitcher01"
17   android:layout_width="wrap_content"
18   android:layout_height="wrap_content" >
19  </ ImageSwitcher>
20/TableLayout>
```

（2）设计控制程序 MainActivity.java。在控制程序 MainActivity.java 中创建图像文件序列数组，并编写按钮的事件处理代码；通过 ViewFactory 接口建立 imageView 图像视图，并实现 OnItemSelectedListener 接口来选择图片。

```java
1 package com.ex2_14;
2 import android.app.Activity;
3 import android.os.Bundle;
4 import android.view.View;
5 import android.view.ViewGroup;
6 import android.view.animation.AnimationUtils;
```

```
7  import android.widget.AdapterView;
8  import android.widget.BaseAdapter;
9  import android.widget.Gallery;
10 import android.widget.ImageSwitcher;
11 import android.widget.ImageView;
12 import android.widget.AdapterView.OnItemSelectedListener;
13 import android.widget.ViewSwitcher.ViewFactory;
14
15 public class MainActivity extends Activity
16 {
17    private ImageSwitcher imageSwitcher;
18     Gallery gallery;
19     private int[] imgs = {
20           R.drawable.img1,
21           R.drawable.img2,
22           R.drawable.img3,
23           R.drawable.img4,
24           R.drawable.img5,
25           R.drawable.img6,
26           R.drawable.img7,
27           R.drawable.img8,
28     };
29
30    @Override
31    public void onCreate(Bundle savedInstanceState)
32    {
33      super.onCreate(savedInstanceState);
34      setContentView(R.layout.activity_main);
35      imageSwitcher = (ImageSwitcher)findViewById(R.id.ImageSwitcher01);
36      imageSwitcher.setFactory(new viewFactory());
37      imageSwitcher.setInAnimation(AnimationUtils
38          .loadAnimation(this, android.R.anim.fade_in) );
39      imageSwitcher.setOutAnimation(AnimationUtils
40          .loadAnimation(this, android.R.anim.fade_out));
41      imageSwitcher.setImageResource(R.drawable.img1);
42      gallery = (Gallery) findViewById(R.id.Gallery01);
43      gallery.setOnItemSelectedListener(
44             new onItemSelectedListener());
45      gallery.setSpacing(10);

46      gallery.setAdapter(new baseAdapter());
47    }

48  //通过ViewFactory接口建立一个imageView图像视图
49  class viewFactory implements ViewFactory
```

- 行37—40 设置淡入（fade_in）、淡出（fade_out）方式
- 行41 设置显示的初始图片
- 行43 设置监听，获取选择的图片
- 行45 设定画廊 gallery 的图片之间的间隔（像素为单位）
- 行46 设置图片文件及显示方式的适配器

```java
50  {
51      @Override
52      public View makeView()         // 在 makeView()中显示视图
53      {
54          ImageView imageView = new ImageView(MainActivity.this);
55          imageView.setScaleType(ImageView.ScaleType.FIT_CENTER)  // 设置显示方式
56          return imageView;
57      }
58  }

59  //实现选项监听接口,获取选择到的图片
60  class onItemSelectedListener  implements OnItemSelectedListener
61  {
62      @Override
63      public void onItemSelected(AdapterView<?> parent,       // 监听选项
64                          View view,int position, long id)
65      {
66          imageSwitcher.setImageResource(
67                  (int)gallery.getItemIdAtPosition(position));
68      }
69      @Override
70      public void onNothingSelected(AdapterView<?> arg0) {    }
71  }

72  //设置一个适配器,安排放在画廊gallery的图片文件及显示方式
73  class  baseAdapter extends BaseAdapter
74  {
75      //取得gallery内的照片数量
76      public int getCount()
77         {return imgs.length;}
78      public Object getItem(int position)
79         { return null;      }
80      //取得gallery内选择的某一张图片文件
81      public long getItemId(int position)
82         { return imgs[position];  }
83      //将选择到的图片放置在imageView,且设定显示方式为居中,大小是60×60
84      public View getView(int position, View convertView, ViewGroup parent)
85      {
86          ImageView imageView = new ImageView(MainActivity.this);
87          imageView.setImageResource(imgs[position]);
88          imageView.setScaleType(ImageView.ScaleType.FIT_CENTER);
89          imageView.setLayoutParams(new Gallery.LayoutParams(60, 60));
90          return imageView;
```

```
91      }
92    }
93 }
```

2.7 消息提示类 Toast

在 Android 系统中可以用 Toast 来显示帮助或提示消息。该提示消息以浮于应用程序之上的形式显示在屏幕上。因为它并不获得焦点，所示不会影响用户的其他操作，使用消息提示组件 Toast 的目的就是为了尽可能不中断用户操作，并使用户看到提供的信息内容。Toast 类的常用属性和对应方法见表 2-15。

表 2-15 Toast 类的常用方法

对应方法	说明
Toast(Context context)	Toast 的构造方法，构造一个空的 Toast 对象
makeText(Context context, CharSequence text, int duration)	以特定时长显示文本内容，参数 text 为显示的文本，参数 duration 为显示的时间，较长时间取值 LENGTH_LONG、较短时间取值 LENGTH_SHORT
getView()	返回视图
setDuration(int duration)	设置存续时间
setView(View view)	设置要显示的视图
setGravity(int gravity, int xOffset, int yOffset)	设置提示信息在屏幕上的显示位置
setText(int resId)	更新 makeText()方法所设置的文本内容
show()	显示提示信息
LENGTH_LONG	提示信息显示较长时间的常量
LENGTH_SHORT	提示信息显示较短时间的常量

【例 2-15】消息提示 Toast 分别按默认方式、自定义方式和带图标方式显示的示例。

将事先准备好的图标文件 icon.jpg 复制到资源目录 res\drawable 下，以做提示消息的图标之用。

（1）设计布局文件 activity_main.xml。在界面设计中设置一个文本标签和 3 个按钮，分别对应消息提示 Toast 的 3 种显示方式，程序代码如下。

```
1    <?xml version="1.0" encoding="utf-8"?>
2    <LinearLayout xmlns:android="http://schemas.android.com/apk/res/android"
3        android:layout_width="fill_parent"
4        android:layout_height="fill_parent"
5        android:orientation="vertical" >
6        <TextView
7            android:layout_width="fill_parent"
8            android:layout_height="wrap_content"
9            android:gravity="center"    ← 居中显示文本
10           android:text="消息提示Toast"
```

```
11          android:textSize="24sp" />
12      <Button
13          android:id="@+id/btn1"
14          android:layout_height="wrap_content"
15          android:layout_width="fill_parent"
16          android:text="默认方式"
17          android:textSize="20sp" />
18      <Button
19          android:id="@+id/btn2"
20          android:layout_height="wrap_content"
21          android:layout_width="fill_parent"
22          android:text="自定义方式"
23          android:textSize="20sp" />
24      <Button
25          android:id="@+id/btn3"
26          android:layout_height="wrap_content"
27          android:layout_width="fill_parent"
28          android:text="带图标方式"
29          android:textSize="20sp" />
30  </LinearLayout>
```

(2) 设计事件处理文件 MainActivity.java。

```
1   package com.ex2_15;
2   import android.app.Activity;
3   import android.os.Bundle;
4   import android.view.Gravity;
5   import android.view.View;
6   import android.view.View.OnClickListener;
7   import android.widget.Button;
8   import android.widget.ImageView;
9   import android.widget.LinearLayout;
10  import android.widget.ListView;
11  import android.widget.Toast;
12
13  public class MainActivity extends Activity
14  {
15      ListView list;
16      Button btn1,btn2,btn3;
17      @Override
18      public void onCreate(Bundle savedInstanceState)
19      {
20          super.onCreate(savedInstanceState);
21          setContentView(R.layout.activity_main);
22          btn1=(Button)findViewById(R.id.btn1);
23          btn2=(Button)findViewById(R.id.btn2);
```

```java
24        btn3=(Button)findViewById(R.id.btn3);
25        btn1.setOnClickListener(new mClick());
26        btn2.setOnClickListener(new mClick());     // 为按钮注册事件监听器
27        btn3.setOnClickListener(new mClick());
28    }
29
30    class mClick implements OnClickListener
31    {
32     Toast toast;
33     LinearLayout toastView;
34     ImageView imageCodeProject;
35     @Override
36     public void onClick(View v)
37     {
38       if(v==btn1)                                 // 居中显示文本
39       {
40         Toast.makeText(getApplicationContext(),   // 设置提示消息内容,可
41                 "默认Toast方式",                    // 用 MainActivity.this 替换
42                 Toast.LENGTH_SHORT).show();        // getApplicationContext()
43       }
44       else if(v==btn2)
45       {
46         toast = Toast.makeText(getApplicationContext(),
47                 "自定义Toast的位置",
48                 Toast.LENGTH_SHORT);
49         toast.setGravity(Gravity.CENTER, 0, 0);    // 自定义显示位置
50         toast.show();
51       }
52       else if(v==btn3)
53       {
54         toast = Toast.makeText(getApplicationContext(),
55                 "带图标的Toast",
56                 Toast.LENGTH_SHORT);
57         toast.setGravity(Gravity.CENTER, 0, 80);
58         toastView = (LinearLayout) toast.getView();       // 定义视图
59         imageCodeProject = new ImageView(MainActivity.this);
60         imageCodeProject.setImageResource(R.drawable.icon);   // 获取资源中的图标
61         toastView.addView(imageCodeProject, 0);           // 在视图中添加图标
62         toast.show();
63       }
64     }
65    }
66 }
```

程序运行结果如图 2.24 所示。

图 2.24 消息提示 Toast 的 3 种方式

2.8 列 表 组 件

2.8.1 列表组件类 ListView

ListView 类是 Android 程序开发中经常用到的组件，该组件必须与适配器配合使用，由适配器提供显示样式和显示数据。

ListView 类的常用属性和对应方法见表 2-16。

表 2-16 ListView 类的常用属性和方法

对应方法	说明
ListView(Context context)	构造方法
setAdapter(ListAdapter adapter)	设置提供数组选项的适配器
addHeaderView(View v)	设置列表项目的头部
addFooterView(View v)	设置列表项目的底部
setOnItemClickListener(AdapterView.OnItemClickListener listener)	注册单击选项时执行的方法，该方法继承于父类 android.widget.AdapterView

【例 2-16】 列表组件示例。

在界面设计中设置一个文本标签和一个列表组件 ListView。

程序设计步骤：

（1）在布局文件中声明列表组件 ListView。

(2) 在 Activity 中获得相关组件实例。
(3) 通过触发列表的选项事件调用 mClick 类的 onClick()方法显示相应提示内容。
程序代码设计：
(1) 设计布局文件 activity_main.xml。

```xml
1  <?xml version="1.0" encoding="utf-8"?>
2  <LinearLayout xmlns:android="http://schemas.android.com/apk/res/android"
3      android:layout_width="fill_parent"
4      android:layout_height="fill_parent"
5      android:orientation="vertical" >
6  <TextView
7      android:layout_width="fill_parent"
8      android:layout_height="wrap_content"
9  android:text="凤凰传奇"
10     android:textSize="24sp" />
11 <ListView
12     android:id="@+id/ListView01"
13     android:layout_height="wrap_content"
14     android:layout_width="fill_parent" />
15 </LinearLayout>
```

(2) 设计事件处理文件 MainActivity.java。

```java
1  package com.ex2_16;
2  import android.app.Activity;
3  import android.os.Bundle;
4  import android.view.View;
5  import android.widget.AdapterView;
6  import android.widget.AdapterView.OnItemClickListener;
7  import android.widget.ArrayAdapter;
8  import android.widget.ListView;
9  import android.widget.TextView;
10 import android.widget.Toast;
11 public class MainActivity extends Activity
12 {
13   ListView list;
14   @Override
15   public void onCreate(Bundle savedInstanceState)
16   {
17     super.onCreate(savedInstanceState);
18     setContentView(R.layout.activity_main);
19     list= (ListView)findViewById(R.id.ListView01);   ← 与界面的列表组件建立关联
20     //定义数组
```

```
21    String[] data ={
22              "(1)荷塘月色",
23              "(2)最炫民族风",
24              "(3)天蓝蓝",
25              "(4)最美天下",
26              "(5)自由飞翔",
27              };
28    //为ListView设置数组适配器ArrayAdapter
29    list.setAdapter(new ArrayAdapter<String>(this,
30       android.R.layout.simple_list_item_1, data));
31    //为ListView设置列表选项监听器
32    list.setOnItemClickListener(new mItemClick());
33  }
34  //定义列表选项监听器的事件
35  class mItemClick implements OnItemClickListener
36  {
37    @Override
38   public void onItemClick(AdapterView<?> arg0, View arg1, int arg2, long arg3)
39    {
40      Toast.makeText(MainActivity,"您选择的项目是："
41        +((TextView)arg1).getText(), Toast.LENGTH_SHORT).show();
42    }
43  }
44 }
```

为列表设置适配器和监听器

提示信息

语句说明：

（1）android.R.layout.simple_list_item_1 是 Android 系统内置的 ListView 布局方式。

- android.R.layout.simple_list_item_1：一行 text。
- android.R.layout.simple_list_item_2：一行 title，一行 text。
- android.R.layout.simple_list_item_single_choice：单选按钮。
- android.R.layout.simple_list_item_multiple_choice：多选按钮。

（2）OnItemClickListener 是一个接口，用于监听列表组件选项的触发事件。

（3）Toast.makeText().show()显示提示消息框。

程序运行结果如图 2.25 所示。

2.8.2　ListActivity 类

当整个 Activity 中只有一个 ListView 组件时可以使用 ListActivity。其实 ListActivity 和只包含一个 ListView 组件的普通 Activity 没有太大的区别，只是实现了一些封装而已。ListActivity 类继承于 Activity 类，默认绑定了一个 ListView 组件，并提供一些与 ListView 处理相关的操作。

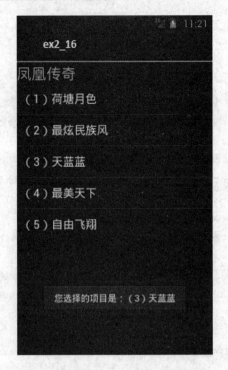

图 2.25 列表组件示例

ListActivity 类常用的方法为 getListView(),该方法返回绑定的 ListView 组件。

【例 2-17】 ListActivity 应用示例。

(1)设计布局文件 activity_main.xml。

```
1 <?xml version="1.0" encoding="utf-8"?>
2 <LinearLayout xmlns:android="http://schemas.android.com/apk/res/android"
3   android:layout_width="fill_parent"
4   android:layout_height="fill_parent"
5   android:orientation="vertical" >
6   <ListView
7     android:id="@+id/android:list"
8     android:layout_height="wrap_content"
9     android:layout_width="fill_parent" />
10 </LinearLayout>
```

说明:ListActivity 的布局文件中的 ListView 组件 id 应设为"@id/android:list"。

(2)设计事件处理文件 MainActivity.java。

```
1 package com.ex2_17;
2 import android.app.ListActivity;
3 import android.os.Bundle;
4 import android.view.View;
```

```java
5  import android.widget.AdapterView;
6  import android.widget.ArrayAdapter;
7  import android.widget.ListView;
8  import android.widget.TextView;
9  import android.widget.Toast;
10 import android.widget.AdapterView.OnItemClickListener;
11 public class MainActivity extends ListActivity
12 {
13   @Override
14   public void onCreate(Bundle savedInstanceState)
15   {
16     super.onCreate(savedInstanceState);
17     setContentView(R.layout.activity_main);
18     //定义数组
19     String[] data ={
20       "(1)荷塘月色",
21       "(2)最炫民族风",
22       "(3)天蓝蓝",
23       "(4)最美天下",
24       "(5)自由飞翔",
25       };
26     //获取列表项
27     ListView list=getListView();
28     //设置列表项的头部
29     TextView header=new TextView(this);
30     header.setText("凤凰传奇经典歌曲");
31     header.setTextSize(24);
32     list.addHeaderView(header);
33     //设置列表项的底部
34     extView foot=new TextView(this);
35     foot.setText("请选择");
36     foot.setTextSize(24);
37     list.addFooterView(foot);
38     setListAdapter(new ArrayAdapter<String>(this,
39       android.R.layout.simple_list_item_1, data));
40     list.setOnItemClickListener(new mItemClick());
41   }
42   //定义列表选项监听器
43   class mItemClick implements OnItemClickListener
44   {
45     @Override
46     public void onItemClick(AdapterView<?> arg0, View arg1, int arg2, long arg3)
```

```
47      {
48          Toast.makeText(getApplicationContext(),
49          "您选择的项目是："+((TextView)arg1).getText(),
50          Toast.LENGTH_SHORT).show();
51      }
52  }
53 }
```

程序运行结果如图 2.26 所示。

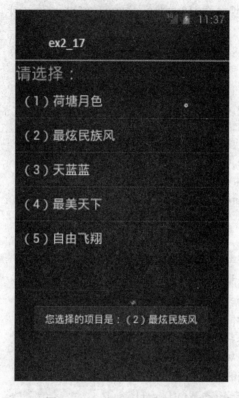

图 2.26   ListActivity 应用示例

2.9 滑动抽屉组件类 SlidingDraw

在日常生活中，当杂乱的物品很多时可以把这些物品分类整理好放在不同的抽屉中，这样在使用物品时打开抽屉里面的东西一目了然。在 Android 系统中也可以把多个程序放到一个应用程序的抽屉里。如图 2.27（a）所示，点击"向上"图标按钮（称为手柄）时打开抽屉；如图 2.27（b）所示，点击"向下"图标按钮时关闭抽屉。

使用 Android 系统提供的 SlidingDraw 组件可以实现滑动抽屉的功能。先来看一下 SlidingDraw 类的重要方法和属性，其重要的 XML 属性如表 2-17 所示。

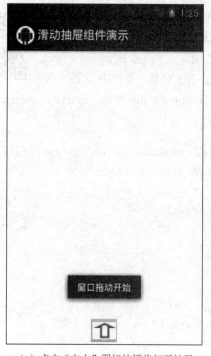 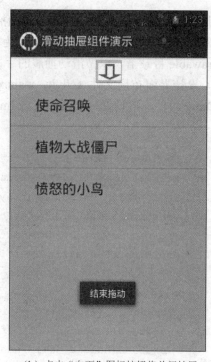

（a）点击"向上"图标按钮将打开抽屉　　　　（b）点击"向下"图标按钮将关闭抽屉

图 2.27　滑动抽屉示例

表 2-17　SlidingDraw 类重要的 XML 属性

属性	说明
android:allowSingleTap	设置通过手柄打开或关闭滑动抽屉
android:animateOnClick	单击手柄时是否加入动画，默认为 true
android:handle	指定抽屉的手柄 handle
android:content	隐藏在抽屉里的内容
android:orientation	滑动抽屉内的对齐方式

SlidingDraw 类的重要方法如表 2-18 所示。

表 2-18　SlidingDraw 类的重要方法

方法	说明
animateOpen()	关闭时实现动画
animateOpen()	打开时实现动画
getContent()	获取内容
getHandle()	获取手柄
setOnDrawerOpenListener(SlidingDrawer.OnDrawerOpenListener onDrawerOpenListener)	打开抽屉的监听器
setOnDrawerCloseListener(SlidingDrawer.OnDrawerCloseListener onDrawerCloseListener)	关闭抽屉的监听器
setOnDrawerScrollListener(SlidingDrawer.OnDrawerScrollListener onDrawerScrollListener)	打开/关闭切换时的监听器

【例 2-18】 实现如图 2.27 所示的滑动抽屉 SlidingDraw 组件应用示例。事先准备好两个图标文件，分别命名为 up.jpg 和 down.jpg，将它们复制到 res\drawable 目录下，以做滑动抽屉的手柄之用。

（1）设计界面布局文件 activity_main.xml。在 XML 文件中设置一个 SlidingDraw 组件，再设置一个图标按钮 ImageButton 做抽屉手柄。activity_main.xml 的代码如下：

```
1   <LinearLayout xmlns:android="http://schemas.android.com/apk/res/android"
2     xmlns:tools="http://schemas.android.com/tools"
3     android:id="@+id/LinearLayout1"
4     android:layout_width="match_parent"
5     android:layout_height="match_parent"
6     android:orientation="vertical" >
7     <!--设置handle和content的id-->
8     <SlidingDrawer
9       android:layout_width="fill_parent"
10      android:layout_height="fill_parent"
11      android:handle="@+id/handle"
12      android:content="@+id/content"
13      android:orientation="vertical"
14      android:id="@+id/slidingdrawer" >
15      <!--设置handle，就是用一个图标按钮来处理滑动抽屉事件-->
16      <ImageButton
17        android:id="@id/handle"
18        android:layout_width="50dip"
19        android:layout_height="44dip"
20        android:src="@drawable/up" />
21      <!--设置抽屉内容，当拖动抽屉的时候就会看到-->
22      <LinearLayout
23        android:id="@id/content"
24        android:layout_width="fill_parent"
25        android:layout_height="fill_parent"
26        android:background="#66cccc"
27        android:focusable="true" >
28      </LinearLayout>
29    </SlidingDrawer>
30  </LinearLayout>
```

（2）设计主控程序 MainActivity.java。在控制程序 MainActivity.java 中主要是实现滑动抽屉的几个监听事件。

```
1   package com.example.ex 2_18;
2   import android.os.Bundle;
```

```java
3   import android.app.Activity;
4   import android.widget.ArrayAdapter;
5   import android.widget.ImageButton;
6   import android.widget.LinearLayout;
7   import android.widget.ListView;
8   import android.widget.SlidingDrawer;
9   import android.widget.Toast;
10
11  public class MainActivity extends Activity
12  {
13      SlidingDrawer mDrawer;
14      ImageButton imgBtn;
15      ListView listView;
16      LinearLayout layout;
17      String data[]=new String[]{"使命召唤","植物大战僵尸","愤怒的小鸟"};
18      @Override
19      public void onCreate(Bundle savedInstanceState)
20      {
21          super.onCreate(savedInstanceState);
22          setContentView(R.layout.activity_main);
23          layout=(LinearLayout) findViewById(R.id.content);
24          listView = new ListView(MainActivity.this);
25          listView.setAdapter(new ArrayAdapter<String>(
26              MainActivity.this,
27              android.R.layout.simple_expandable_list_item_1,
28              data));
29          layout.addView(listView);
30          imgBtn=(ImageButton)findViewById(R.id.handle);
31          mDrawer=(SlidingDrawer)findViewById(R.id.slidingdrawer);
32          mDrawer.setOnDrawerOpenListener(new mOpenListener());
33          mDrawer.setOnDrawerCloseListener(new mCloseListener());
34          mDrawer.setOnDrawerScrollListener(new mScrollListener());
35      }
36
37      class mOpenListener implements SlidingDrawer.OnDrawerOpenListener
38      {
39          @Override
40          public void onDrawerOpened()
41          {
42              imgBtn.setImageResource(R.drawable.down);
43          }
44      }
45
```

注释:
- 在抽屉布局中创建一个视图,显示数组内容 (对应第23–28行)
- 注册监听器 (对应第31–34行)
- 打开抽屉时触发 (对应第37–44行)

```
46    class mCloseListener implements SlidingDrawer.OnDrawerCloseListener
47    {
48      @Override
49      public void onDrawerClosed()
50      {
51            imgBtn.setImageResource(R.drawable.up);
52      }
53    }
54
55    class mScrollListener implements SlidingDrawer.OnDrawerScrollListener
56    {
57      @Override
58      public void onScrollEnded()
59      {
60        Toast.makeText(MainActivity.this, "结束拖动",
61                    Toast.LENGTH_SHORT).show();
62      }
63      @Override
64      public void onScrollStarted()
65      {
66        Toast.makeText(MainActivity.this, "窗口拖动开始",
67                    Toast.LENGTH_SHORT).show();
68      }
69    }
70  }
```

关闭抽屉时触发

打开/关闭切换时触发

习 题 2

1. 编写程序，实现点击按钮将文本编辑框中输入的文字内容显示到文本标签，如图 2.28 所示。

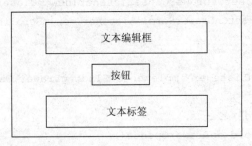

图 2.28 文本编辑框中的文字内容显示到文本标签

2. 设计一个加法计算器，如图 2.29 所示，在前两个文本编辑框中输入整数，当点击按钮"="时在第 3 个文本编辑框中显示这两个数之和。

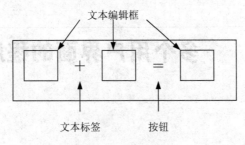

图 2.29 加法计算器

3. 设计如图 2.30 所示的用户界面布局。

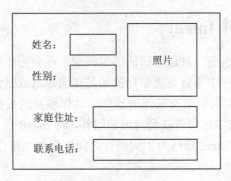

图 2.30 设计用户界面

4. 完成例 2-9 计算器的 Java 控制程序代码设计。

5. 编写一个"我的故乡"图册，下方配有说明文字，点击"上一张"或"下一张"按钮，图片切换时说明文字内容也随之切换。

第 3 章　多个用户界面的程序设计

3.1　页面的切换与传递参数值

3.1.1　传递参数组件 Intent

Intent 是 Android 系统的一种运行时的绑定机制，在应用程序运行时连接两个不同组件。在 Android 的应用程序中不管是页面切换还是传递数据或是调用外部程序都可能要用到 Intent。Intent 负责对应用中某次操作的动作、动作涉及的数据、附加数据进行描述，Android 则根据此 Intent 的描述负责找到对应的组件，将 Intent 传递给调用的组件，并完成组件的调用。因此可以将 Intent 理解为不同组件之间通信的"媒介"，其专门提供组件互相调用的相关信息。

Intent 的属性有动作（Action）、数据（Data）、分类（Category）、类型（Type）、组件（Compent）以及扩展（Extra），其中最常用的是 Action 属性。

例如：

Intent.ACTION_MAIN	表示标识 Activity 为一个程序的开始。
Intent.ACTION_GET_CONTENT	表示允许用户选择图片或录音等特殊种类的数据。
Intent.ACTION_SEND	表示发送邮件的 action 动作。
Telephony.SMS_RECEIVED	表示接收邮件的 action 动作。
Intent.ACTION_ANSWER	表示处理呼入的电话。
Intent.Action_CALL_BUTTON	表示按"拨号"键。
Intent.Action_CALL	表示呼叫指定的电话号码。

3.1.2　Activity 页面的切换

Activity 跳转与传递参数值主要通过 Intent 类协助实现。在一个 Activity 页面中启动另一个 Activity 页面的运行是最简单的 Activity 页面切换方式。其步骤如下：

（1）创建一个 Intent 对象，其构造方法如下。

```
Intent intent = new Intent(当前Activity.this, 另一Activity.class);
```

（2）调用 Activity 的 startActivity(intent)方法，切换到另一个 Activity 页面。

【例 3-1】　从一个 Activity 页面启动另一个 Activity 页面示例。

创建名称为 ex3_1 的新项目，包名为 com.ex3_1。在本项目中要建立两个页面文件及两个控制文件，第 1 个页面的界面布局文件为 activity_main.xml、控制文件为 MainActivity.java，第 2 个页面的界面布局文件为 second.xml、控制文件为 secondActivity.java，还要修改配置文件 AndroidManifest.xml。

（1）设计第 1 个页面。

① 修改第 1 个页面的控制文件 MainActivity.java，源代码如下：

```
1  package com.ex3_1;
2  import android.app.Activity;
3  import android.content.Intent;
4  import android.os.Bundle;
5  import android.view.View;
6  import android.view.View.OnClickListener;
7  import android.widget.Button;
8  public class MainActivity extends Activity
9  {
10    private Button btn;
11    @Override
12    public void onCreate(Bundle savedInstanceState)
13    {
14     super.onCreate(savedInstanceState);
15     setContentView(R.layout.activity_main);
16     btn = (Button)findViewById(R.id.mButton);
17     btn.setOnClickListener(new btnclock());
18    }
19   class btnclock implements OnClickListener     ← 定义一个类实现监听接口
20   {
21    public void onClick(View v)
22    {
23      Intent intent = new Intent(MainActivity.this, secondActivity.class);
24       //创建好Intent之后就可以通过它启动新的Activity
25      startActivity(intent);
26    }
27   }
28  }
```

② 第 1 个页面的布局文件 activity_main.xml 如下：

```
1  <?xml version="1.0" encoding="utf-8"?>
2  <LinearLayout xmlns:android="http://schemas.android.com/apk/res/android"
3    android:layout_width="fill_parent"
4    android:layout_height="fill_parent"
5    android:orientation="vertical" >
6    <TextView
7      android:id="@+id/textView1"
8      android:layout_width="fill_parent"
9      android:layout_height="wrap_content"
```

```
10    android:text="@string/hello" />
11  <Button
12    android:id="@+id/mButton"
13    android:layout_width="wrap_content"
14    android:layout_height="wrap_content"
15    android:text="@string/button"
16    />
17 </LinearLayout>
```

(2) 设计第 2 个页面。

① 在项目中新建第 2 个页面的控制文件 secondActivity.java。右击资源管理器中的 com.example.ex3_1 选项，在弹出的快捷菜单中选择 New（新建）→File（文件）命令，如图 3.1 所示。

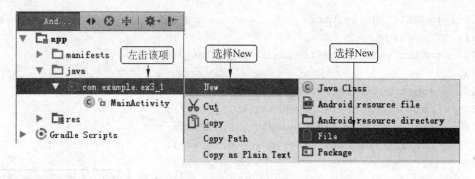

图 3.1　新建一个 Java 源程序

在弹出的对话框中输入文件名"secondActivity.java"，并输入代码如下：

```
1 package com.ex3_1;
2 import android.app.Activity;
3 import android.os.Bundle;
4 public class secondActivity extends Activity
5 {
6   @Override
7   public void onCreate(Bundle savedInstanceState)
8   {
9     super.onCreate(savedInstanceState);
10    setContentView(R.layout.second);   ← 启动布局文件 second.xml
11  }
12 }
```

② 新建第 2 个页面的布局文件 second.xml。其操作同前，右击资源管理器中的 layout 选项，在弹出的快捷菜单中选择 New（新建）→File（文件）命令，然后在弹出的对话框中输入文件名"second.xml"，并输入其代码如下。

```
1 <?xml version="1.0" encoding="utf-8"?>
```

```
2  <LinearLayout xmlns:android="http://schemas.android.com/apk/res/android"
3    android:layout_width="fill_parent"
4    android:layout_height="fill_parent"
5    android:orientation="vertical" >
6    <TextView
7      android:layout_width="fill_parent"
8      android:layout_height="wrap_content"
9      android:text="@string/second" />
10 </LinearLayout>
```

（3）修改 strings.xml 和配置文件 AndroidManifest.xml。

① strings.xml 文件的代码如下：

```
1 <?xml version="1.0" encoding="utf-8"?>
2 <resources>
3   <string name="hello">切换页面</string>
4   <string name="app_name">ex3_1</string>
5   <string name="second"> 这是第2个页面  </string>
6   <string name="button">切换到另一页面</string>
7 </resources>
```

② 修改 AndroidManifest.xml 配置文件。打开项目中的 AndroidManifest.xml 文件，向其注册第 2 个 Activity 页面，其代码如下：

```
1  <?xml version="1.0" encoding="utf-8"?>
2  <manifest xmlns:android="http://schemas.android.com/apk/res/android"
3    package="com.ex3_1"
4    android:versionCode="1"
5    android:versionName="1.0" >
6    <uses-sdk android:minSdkVersion="15" />
7    <application
8      android:icon="@drawable/ic_launcher"
9      android:label="@string/app_name" >
10     <activity
11       android:label="@string/app_name"
12       android:name=".MainActivity" >
13       <intent-filter >
14         <action android:name="android.intent.action.MAIN" />
15         <category android:name="android.intent.category.LAUNCHER" />
16       </intent-filter>
17     </activity>
18     <activity
19       android:label="@string/app_name"
20       android:name=".secondActivity" />
```

新增第 2 个 Activity 的注册

```
21    </application>
22 </manifest>
```

程序运行结果如图 3.2 所示。

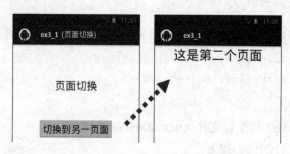

图 3.2 从一个页面切换到另一页面

提示：在 Android Studio 中创建第 2 个页面也可以通过下列操作由系统自动生成：右击资源管理器中应用程序的 app 项，选择 New→File→Activity→Gallery 项，根据系统对话框的导航提示自动创建第 2 个页面（包括控制程序、界面布局程序，并在配置文件中自动添加第 2 个页面的 Activity 注册语句）。

3.1.3 应用 Intent 在 Activity 页面之间传递数据

1. Bundle 类

Bundle 类是用于为字符串与某组件对象建立映射关系的组件。Bundle 组件与 Intent 配合使用，可在不同的 Activity 之间传递数据。Bundle 类的常用方法如下。

- putString(String key, String value)：把字符串用"键－值"对形式存放到 Bundle 对象中。
- remove(String key)：移除指定 key 的值。
- getString(String key)：获取指定 key 的字符。

2. 应用 Intent 在不同的 Activity 之间传递数据

下面说明应用 Intent 与 Bundle 配合从一个 Activity 页面传递数据到另一 Activity 页面的方法。

1）在页面 Activity A 端

（1）创建 Intent 对象和 Bundle 对象：

```
Intent intent = new Intent();
Bundle bundle = new Bundle();
```

（2）为 Intent 指定切换页面，用 Bundle 存放"键－值"对数据：

```
intent.setClass(MainActivity.this, secondActivity.class);
bundle.putString("text", txt.getText().toString());
```

（3）将 Bundle 对象传递给 Intent：

```
intent.putExtras(bundle);
```

2) 在另一页面 Activity B 端
（1）从 Intent 中获取 Bundle 对象：

```
bunde = this.getIntent().getExtras();
```

（2）从 Bundle 对象中按"键-值"对的键名获取对应数据值：

```
String str = bunde.getString("text");
```

在不同的 Activity 页面之间传递数据的过程如图 3.3 所示。

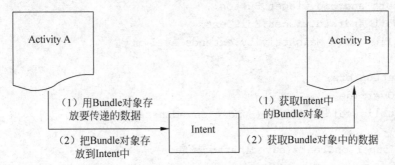

图 3.3　应用 Intent 在 Activity 页面之间传递数据

【例 3-2】　从第 1 个 Activity 页面传递数据到第 2 个 Activity 页面示例。

（1）第 1 个 Activity 页面的界面布局 activity_main.xml 的代码如下：

```
1  <?xml version="1.0" encoding="utf-8"?>
2  <LinearLayout xmlns:android="http://schemas.android.com/apk/res/android"
3      android:layout_width="fill_parent"
4      android:layout_height="fill_parent"
5      android:orientation="vertical" >
6      <TextView
7          android:layout_width="fill_parent"
8          android:layout_height="wrap_content"
9          android:text="页面切换" />
10     <Button
11         android:id="@+id/button1"
12         android:layout_width="wrap_content"
13         android:layout_height="wrap_content"
14         android:text="切换到另一页面" />
15     <EditText
16         android:id="@+id/editText1"
17         android:layout_width="match_parent"
18         android:layout_height="wrap_content" >
19         <requestFocus />      ← 获得焦点
```

```
20   </EditText>
21 </LinearLayout>
```

(2)第 1 个 Activity 页面控制文件 MainActivity.java 的代码如下:

```java
1  package com.ex3_2;
2  import android.app.Activity;
3  import android.content.Intent;
4  import android.os.Bundle;
5  import android.view.View;
6  import android.view.View.OnClickListener;
7  import android.widget.Button;
8  import android.widget.EditText;
9  public class MainActivity extends Activity
10 {
11   Button btn;
12   @Override
13   public void onCreate(Bundle savedInstanceState)
14   {
15     super.onCreate(savedInstanceState);
16     setContentView(R.layout.activity_main);
17     btn = (Button)findViewById(R.id.button1);
18     btn.setOnClickListener(new btnclock());
19   }
20   //定义一个类实现监听接口
21   class btnclock implements OnClickListener
22   {
23    public void onClick(View v)
24    {
25     Intent intent = new Intent();              // 创建 Intent 对象并指定切换页面
26     intent.setClass(mainActivity.this,secondActivity.class);
27     EditText txt = (EditText)findViewById(R.id.editText1);
28     Bundle bundle = new Bundle();              // 创建 Bundle 对象存放"键—值"对数据
29     bundle.putString("text", txt.getText().toString());
30     intent.putExtras(bundle);                  // 将 Bundle 对象传递给 Intent
31     startActivity(intent);                     // 启动另一个 Activity 页面
32    }
33   }
34 }
```

(3)第 2 个页面的界面布局文件 second.xml 的代码如下:

```xml
1 <?xml version="1.0" encoding="utf-8"?>
2 <LinearLayout xmlns:android="http://schemas.android.com/apk/res/android"
```

```
3      android:layout_width="fill_parent"
4      android:layout_height="fill_parent"
5      android:orientation="vertical" >
6      <Button
7        android:id="@+id/button2"
8        android:layout_width="wrap_content"
9        android:layout_height="wrap_content"
10       android:text="返回第一个页面" />
11     <TextView
12       android:id="@+id/TextView2"
13       android:layout_width="match_parent"
14       android:layout_height="wrap_content"
15       android:textSize="24sp"    />
16  </LinearLayout>
```

（4）第 2 个页面的控制文件 secondActivity.java 的代码如下：

```
1   package com.ex3_2;
2   import android.app.Activity;
3   import android.content.Intent;
4   import android.os.Bundle;
5   import android.util.Log;
6   import android.view.View;
7   import android.view.View.OnClickListener;
8   import android.widget.Button;
9   import android.widget.TextView;
10  public class secondActivity extends Activity
11  {
12    Button btn2;
13    @Override
14   public void onCreate(Bundle savedInstanceState)
15    {
16     super.onCreate(savedInstanceState);
17     setContentView(R.layout.second);
18     TextView txt2 = (TextView)findViewById(R.id.TextView2);
19     Bundle bunde = this.getIntent().getExtras();      ← 取得 Intent 中的 Bundle 对象
20
21     String str = bunde.getString("text");      ← 获取 Bundle 对象中的数据
22     txt2.setText(str);
23     btn2 = (Button)findViewById(R.id.button2);
24     btn2.setOnClickListener(new btnclock2() );
25    }
26   //定义返回到前一页面的监听接口事件
27   class btnclock2 implements OnClickListener
28   {
29     public void onClick(View v)
```

```
30      {
31          Intent intent2 = new Intent();      ← 新建 Intent 对象
32          intent2.setClass(secondActivity.this, MainActivity.class);
33          startActivityForResult(intent2, 0); ← 返回前一页面
34      }
35   }
36 }
```

（5）修改 AndroidManifest.xml 配置文件。打开项目中的 AndroidManifest.xml 文件，向其注册第 2 个 Activity 页面，其代码同例 3-1。

程序运行结果如图 3.4 所示。

图 3.4 数据在不同 Activity 页面之间传递

3.2 菜单设计

一个菜单（Menu）由多个菜单选项组成，选择一个菜单项就可以引发一个动作事件。

在 Android 系统中，菜单可以分为 3 类，即选项菜单（Option Menu）、上下文菜单（Context Menu）和子菜单（Sub Menu）。下面主要介绍选项菜单和上下文菜单的设计方法，由于子菜单的设计方法基本上与选项菜单相同，这里就不赘述了。

3.2.1 选项菜单

选项菜单需要通过按下设备的 Menu 键来显示。当按下设备上的 Menu 键后会在屏幕底部弹出一个菜单，这个菜单称为选项菜单（Options Menu）。

1．在 Activity 中创建菜单的方法

设计选项菜单需要用到 Activity 中的 onCreateOptionMenu(Menu menu)方法，用于建立菜单并且在菜单中添加菜单项；还需要用到 Activity 中的 onOptionsItemSelected(MenuItem item)方法，用于响应菜单事件。Activity 实现选项菜单的方法见表 3-1。

表 3-1 Activity 实现选项菜单的方法

方法	说明
onCreateOptionMenu(Menu menu)	用于初始化菜单，menu 为 Menu 对象实例
onPrepareOptionsMenu(Menu menu)	改变菜单状态，在菜单显示前调用
onOptionsMenuClosed(Menu menu)	菜单被关闭时调用
onOptionsItemSelected(MenuItem item)	菜单项被点击时调用，即菜单项的监听方法

2. 菜单 Menu

设计选项菜单需要用到 Menu、MenuItem 接口。一个 Menu 对象代表一个菜单，在 Menu 对象中可以添加菜单项 MenuItem 对象，也可以添加子菜单 Sub Menu。

菜单 Menu 使用 add(int groupId, int itemId, int order, CharSequence title) 方法添加一个菜单项，add()方法中的 4 个参数如下。

（1）组别：如果不分组就写 Menu.NONE。

（2）id：很重要，Android 根据这个 id 来确定不同的菜单。

（3）顺序：哪个菜单项在前面由这个参数的大小决定。

（4）文本：菜单项的显示文本。

3. 创建选项菜单的步骤

创建选项菜单的步骤如下：

（1）重写 Activity 的 onCreateOptionMenu(Menu menu)方法，当菜单第 1 次被打开时调用。

（2）调用 Menu 的 add()方法添加菜单项（MenuItem）。

（3）重写 Activity 的 onOptionsItemSelected(MenuItem item)方法，当菜单项（MenuItem）被选择时来响应事件。

【例 3-3】 选项菜单应用示例。

设计一个应用选项菜单的示例程序，其运行结果如图 3.5 所示。

图 3.5 菜单示例

其 MainActivity.java 的源代码如下：

```
1 package com.ex3_3;
2 import android.app.Activity;
```

```
3  import android.os.Bundle;
4  import android.view.Menu;
5  import android.view.MenuItem;
6  import android.widget.TextView;
7  public class MainActivity extends Activity
8  {
9    TextView txt;
10    @Override
11    public void onCreate(Bundle savedInstanceState)
12    {
13        super.onCreate(savedInstanceState);
14        setContentView(R.layout.activity_main);
15      txt = (TextView)findViewById(R.id.TextView1);
16    }
17    @Override
18    public boolean onCreateOptionsMenu(Menu menu)
19    {
20        //调用父类方法来加入系统菜单
21      super.onCreateOptionsMenu(menu);
22        //添加菜单项
23      menu.add(
24                1,          //组号
25                1,          //唯一的id号
26                1,          //排序号
27                "菜单项1");  //标题
28      menu.add( 1, 2, 2, "菜单项2");
29      menu.add( 1, 3, 3, "菜单项3");
30      menu.add( 1, 4, 4, "菜单项4");
31      return true;
32    }
33     @Override
34    public boolean onOptionsItemSelected(MenuItem item)
35    {
36      String title = "选择了" + item.getTitle().toString();
37      switch (item.getItemId())
38      { //响应每个菜单项（通过菜单项的id）
39      case 1:
40          txt.setText(title);    ← 文本标签显示菜单项的标题
41          break;
42      case 2:
43          txt.setText(title);    ← 文本标签显示菜单项的标题
44          break;
45      case 3:
46          txt.setText(title);    ← 文本标签显示菜单项的标题
47          break;
```

添加菜单项的 4 个参数

```
48        case 4:
49            txt.setText(title);    ← 文本标签显示菜单项的标题
50            break;
51        default:
52            //对于没有处理的事件交给父类来处理
53            return super.onOptionsItemSelected(item);
54        }
55        return true;
56    }
57 }
```

3.2.2 上下文菜单

Android 系统中的上下文菜单类似于计算机上的右键菜单。在为一个视图注册了上下文菜单之后长按（两秒左右）这个视图对象会弹出一个浮动菜单，即上下文菜单。任何视图都可以注册上下文菜单，不过最常见的是用于列表视图 ListView 的 item。

创建一个上下文菜单的步骤如下：

（1）重写 Activity 的 onCreateContenxtMenu()方法，调用 Menu 的 add 方法添加菜单项（MenuItem）。

（2）重写 Activity 的 onContextItemSelected()方法，响应上下文菜单的菜单项的单击事件。

（3）调用 Activity 的 registerForContextMenu()方法，为视图注册上下文菜单。

【例 3-4】 上下文菜单应用示例。

设计一个应用上下文菜单的示例程序，其运行结果如图 3.6 所示。

图 3.6　上下文菜单应用示例

其 MainActivity.java 的源代码如下：

```
1  package com.ex3_4;
2  import android.app.Activity;
```

```
3   import android.os.Bundle;
4   import android.view.ContextMenu;
5   import android.view.ContextMenu.ContextMenuInfo;
6   import android.view.Menu;
7   import android.view.MenuItem;
8   import android.view.View;
9   import android.widget.ListView;
10  import android.widget.TextView;
11  public class MainActivity extends Activity
12  {
13    TextView txt1, txt2, txt3;
14    private static final int item1 = Menu.FIRST;
15    private static final int item2 = Menu.FIRST+1;
16    private static final int item3 = Menu.FIRST+2;
17    String str[] = {" 令狐冲","杨   过","萧   峰 " };
18    @Override
19    public void onCreate(Bundle savedInstanceState)
20    {
21      super.onCreate(savedInstanceState);
22      setContentView(R.layout.activity_main);
23      txt1=(TextView)findViewById(R.id.textView1);
24      txt2=(TextView)findViewById(R.id.textView2);
25      txt3=(TextView)findViewById(R.id.textView3);
26      txt1.setText(str[0].toString());
27      txt2.setText(str[1].toString());
28      txt3.setText(str[2].toString());
29      registerForContextMenu(txt1);
30      registerForContextMenu(txt2);
31      registerForContextMenu(txt3);
32    }
33    //上下文菜单，本例会通过长按条目激活上下文菜单
34    @Override
35    public void onCreateContextMenu(ContextMenu menu, View view,
36    ContextMenuInfo menuInfo) {
37        menu.setHeaderTitle("人物简介");
38        //添加菜单项
39        menu.add(0, item1, 0, "武功");
40        menu.add(0, item2, 0, "战斗力");
41        menu.add(0, item3, 0, "经典语录");
42    }
43    //菜单单击响应
44    @Override
45    public boolean onContextItemSelected(MenuItem item){
46        //获取当前被选择的菜单项的信息
47        switch(item.getItemId())
```

（第 23-31 行：初始化组件）

```
48    {
49        case item1:
50            //在这里添加处理代码
51            break;
52        case item2:
53            //在这里添加处理代码
54            break;
55        case item3:
56            //在这里添加处理代码
57            break;
58    }
59    return true;
60  }
61 }
```

← 选项的具体功能没有实现

3.3 对 话 框

对话框是一个有边框、有标题栏的独立存在的容器,在应用程序中经常使用对话框组件来进行人机交互。Android 系统提供了 4 种常用的对话框。

- AlertDialog:消息对话框。
- ProgressDialog:进度条对话框。
- DatePickerDialog:日期选择对话框。
- TimePickerDialog:时间选择对话框。

下面逐一介绍这些对话框的使用方法。

3.3.1 消息对话框 AlertDialog

AlertDialog 对话框是应用程序设计中最常用的对话框之一。AlertDialog 对话框的内容很丰富,使用 AlertDialog 可以创建普通对话框、带列表的对话框以及带单选按钮和多选按钮的对话框。AlertDialog 的常用方法如表 3-2 所示。

表 3-2 AlertDialog 的常用方法

方法	说明
AlertDialog.Builder(Context)	对话框 Builder 对象的构造方法
create();	创建 AlertDialog 对象
setTitle();	设置对话框的标题
setIcon();	设置对话框的图标
setMessage();	设置对话框的提示信息
setItems();	设置对话框要显示的一个 list
setPositiveButton();	在对话框中添加 yes 按钮
setNegativeButton();	在对话框中添加 no 按钮
show();	显示对话框
dismiss();	关闭对话框

创建 AlertDialog 对象需要使用 AlertDialog 的内部类 Builder。设计 AlertDialog 对话框的步骤如下：

（1）用 AlertDialog.Builder 类创建对话框 Builder 对象：

```
Builder dialog=new AlertDialog.Builder(Context);
```

（2）设置对话框的标题、图标、提示信息内容、按钮等：

```
dialog.setTitle("普通对话框");
dialog.setIcon(R.drawable.icon1);
dialog.setMessage("一个简单的提示对话框");
dialog.setPositiveButton("确定", new okClick());
```

（3）创建并显示 AlertDialog 对话框对象：

```
dialog.create();
dialog.show();
```

如果在对话框内部设置了按钮，还需要为其设置事件监听 OnClickListener。

【例 3-5】 消息对话框应用示例。

在本例中设计了两种形式的对话框程序，一种是发出提示信息的普通对话框，另一种是用户登录对话框。

在用户登录对话框中设计了用户登录的布局文件 long.xml 供用户输入相关验证信息。

程序的运行结果如图 3.7 所示。

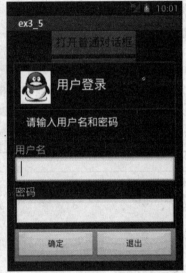

（a）普通对话框　　　　　　　　　（b）用户登录对话框

图 3.7　AlertDialog 对话框

（1）设计界面布局文件 activity_main.xml。

```
1    <?xml version="1.0" encoding="utf-8"?>
```

```
2   <LinearLayout xmlns:android="http://schemas.android.com/apk/res/android"
3       android:layout_width="fill_parent"
4       android:layout_height="fill_parent"
5       android:gravity="center_horizontal"
6       android:orientation="vertical" >
7       <Button
8           android:id="@+id/button1"
9           android:layout_width="wrap_content"
10          android:layout_height="wrap_content"
11          android:text="打开普通对话框"
12          android:textSize="20sp"
13      />
14      <Button
15          android:id="@+id/button2"
16          android:layout_width="wrap_content"
17          android:layout_height="wrap_content"
18          android:text="打开输入对话框"
19          android:textSize="20sp"
20      />
21  </LinearLayout>
```

（2）设计登录对话框的界面布局文件 login.xml。

```
1   <?xml version="1.0" encoding="utf-8"?>
2   <LinearLayout xmlns:android="http://schemas.android.com/apk/res/android"
3       android:layout_width="fill_parent"
4       android:layout_height="fill_parent"
5       android:orientation="vertical" >
6       <TextView
7           android:id="@+id/user"
8           android:layout_width="fill_parent"
9           android:layout_height="wrap_content"
10          android:text="用户名"
11          android:textSize="18sp"/>
12      <EditText
13          android:id="@+id/userEdit"
14          android:layout_width="fill_parent"
15          android:layout_height="wrap_content"
16          android:textSize="18sp"/>
17      <TextView
18          android:id="@+id/password"
19          android:layout_width="fill_parent"
20          android:layout_height="wrap_content"
21          android:text="密码"
22          android:textSize="18sp"/>
23      <EditText
```

```
24            android:id="@+id/paswdEdit"
25            android:layout_width="fill_parent"
26            android:layout_height="wrap_content"
27            android:textSize="18sp"/>
28    </LinearLayout>
```

(3) 设计控制文件 MainActivity.java。

```java
1   package com.ex3_5;
2   import android.app.Activity;
3   import android.app.AlertDialog;
4   import android.app.AlertDialog.Builder;
5   import android.app.ProgressDialog;
6   import android.content.DialogInterface;
7   import android.os.Bundle;
8   import android.view.View;
9   import android.view.View.OnClickListener;
10  import android.widget.Button;
11  import android.widget.EditText;
12  import android.widget.LinearLayout;
13  import android.widget.Toast;
14
15  public class MainActivity extends Activity
16  {
17    ProgressDialog mydialog;
18    Button btn1,btn2;
19    LinearLayout login;
20    @Override
21    public void onCreate(Bundle savedInstanceState)
22    {
23       super.onCreate(savedInstanceState);
24       setContentView(R.layout.activity_main);
25       btn1=(Button)findViewById(R.id.button1);
26       btn2=(Button)findViewById(R.id.button2);
27       btn1.setOnClickListener(new mClick());
28       btn2.setOnClickListener(new mClick());
29    }
30    class mClick implements OnClickListener
31    {
32    Builder dialog=new AlertDialog.Builder(MainActivity.this);
33    @Override
34    public void onClick(View arg0)
35    {
36     if(arg0 == btn1)
37      {
```

```
38       //设置对话框的标题
39       dialog.setTitle("警告");
40       //设置对话框的图标
41       dialog.setIcon(R.drawable.icon1);               ← 设置对话框
42       //设置对话框显示的内容
43       dialog.setMessage("本项操作可能导致信息泄漏！");
44       //设置对话框的"确定"按钮
45       dialog.setPositiveButton("确定", new okClick());
46       //创建对象框
47       dialog.create();
48       //显示对象框
49       dialog.show();
50     }
51     else  if(arg0 == btn2)
52     {
53      login = (LinearLayout)getLayoutInflater()
54             .inflate(R.layout.login, null);          ← 从另外的布局中关联组件
55     dialog.setTitle("用户登录").setMessage("请输入用户名和密码")
56             .setView(login);
57     dialog.setPositiveButton("确定", new loginClick());
58     dialog.setNegativeButton("退出", new exitClick());
59     dialog.setIcon(R.drawable.icon2);
60      dialog.create();
61      dialog.show();
62     }
63    }
64  }
65   /*普通对话框的"确定"按钮事件*/
66   class okClick implements DialogInterface.OnClickListener
67   {
68    @Override
69    public void onClick(DialogInterface dialog, int which)
70    {
71       dialog.cancel();   ← 关闭对话框
72    }
73  }
74   /*输入对话框的"确定"按钮事件*/
75   class loginClick implements  DialogInterface.OnClickListener
76   {
77    EditText txt;
78     @Override
79    public void onClick(DialogInterface dialog, int which)
80    {
81      txt = (EditText)login.findViewById(R.id.paswdEdit);  ← 关联布局文件中的组件
82       //取出输入编辑框的值与密码"admin"比较
```

```
83      if((txt.getText().toString()).equals("admin"))
84          Toast.makeText(getApplicationContext(),
85              "登录成功", Toast.LENGTH_SHORT).show();
86      else
87          Toast.makeText(getApplicationContext(),
88              "密码错误", Toast.LENGTH_SHORT).show();
89      dialog.dismiss();   ← 关闭对话框
90      }
91   }
92   /*  输入对话框的"退出"按钮事件   */
93   class exitClick implements DialogInterface.OnClickListener
94   {
95      @Override
96      public void onClick(DialogInterface dialog, int which)
97      {
98          MainActivity.this.finish();
99      }
100  }
101 }
```

密码为 admin 时显示"登录成功"

点击"退出"按钮退出 MainActivity 程序

对于程序的第 53、54 行：

```
login = (LinearLayout)getLayoutInflater()
            .inflate(R.layout.login, null);
```

这里 inflate 是将组件从一个 XML 中定义的布局找出来。

在一个 Activity 中如果直接用 findViewById()，对应的是 setConentView()中的那个 layout 中的组件（程序第 24 行中的 R.layout.activity_main）。如果 Activity 中用到其他 layout 布局，比如对话框上的 layout,还要设置对话框上的 layout 中的组件（像图片 ImageView、文字 TextView）上的内容，这就必须用 inflate()先将对话框上的 layout 找出来,然后再用这个 layout 对象找到它上面的组件。

3.3.2 其他几种常用对话框

1. 进度条对话框 ProgressDialog

Android 系统有一个 ProgressDialog 类，它继承于 AlertDialog，综合了进度条与对话框的特点，使用起来非常简单。ProgressDialog 类的继承关系如图 3.7 所示。

```
java.lang.Object
   └ android.app.Dialog
       └ android.app.AlertDialog
           └ android.app.ProgressDialog
```

图 3.7　ProgressDialog 类继承于 AlertDialog

ProgressDialog 的常用方法见表 3-3。

表 3-3　ProgressDialog 的常用方法

方法	说明
getMax()	获取对话框进度的最大值
getProgress()	获取对话框当前的进度值
onStart()	开始调用对话框
setMax(int max)	设置对话框进度的最大值
setMessage(CharSequence message)	设置对话框的文本内容
setProgress(int value)	设置对话框当前的进度
show(Context context, CharSequence title, CharSequence message)	设置对话框的显示内容和方式
ProgressDialog(Context context)	对话框的构造方法

2. 日期选择对话框和时间选择对话框

日期选择对话框 DatePickerDialog 和时间选择对话框 TimePickerDialog 都继承于 AlertDialog，一般用于日期和时间的设定，它们的常用方法如表 3-4 所示。

表 3-4　日期和时间选择对话框的常用方法

方法	说明
updateDate(int year, int monthOfYear, int dayOfMonth)	设置 DatePickerDialog 对象的当前日期
onDateChanged(DatePicker view, int year, int month, int day)	修改 DatePickerDialog 对象的日期
updateTime(int hourOfDay, int minuteOfHour)	设置 TimePickerDialog 对象的时间
onTimeChanged(TimePicker view, int hourOfDay, int minute)	修改 TimePickerDialog 对象的时间

【例 3-6】 进度及日期、时间对话框示例。

```
1   package com.example.ex3_6;
2   import android.app.Activity;
3   import android.app.DatePickerDialog;
4   import android.app.ProgressDialog;
5   import android.app.TimePickerDialog;
6   import android.app.DatePickerDialog.OnDateSetListener;
7   import android.app.TimePickerDialog.OnTimeSetListener;
8   import android.os.Bundle;
9   import android.view.View;
10  import android.view.View.OnClickListener;
11  import android.widget.Button;
12  import android.widget.DatePicker;
13  import android.widget.TimePicker;
14
15  public class MainActivity extends Activity
16  {
17      Button btn1,btn2,btn3;
18      @Override
19      public void onCreate(Bundle savedInstanceState)
```

```java
20  {
21      super.onCreate(savedInstanceState);
22      setContentView(R.layout.activity_main);
23      btn1=(Button)findViewById(R.id.button1);
24      btn2=(Button)findViewById(R.id.button2);
25      btn3=(Button)findViewById(R.id.button3);
26      btn1.setOnClickListener(new mClick());
27      btn2.setOnClickListener(new mClick());
28      btn3.setOnClickListener(new mClick());
29  }
30  class mClick implements OnClickListener
31  {
32      int m_year = 2012;
33      int m_month = 1;
34      int m_day = 1;
35      int m_hour = 12,  m_minute = 1;
36      @Override
37      public void onClick(View v)
38      {
39          if(v == btn1)
40          {
41              ProgressDialog d=new ProgressDialog (MainActivity.this);
42              d.setTitle("进度对话框");
43              d.setIndeterminate(true);
44              d.setMessage("程序正在Loading...");
45              d.setCancelable(true);
46              d.setMax(10);
47              d.show();
48          }
49          else if(v == btn2)
50          {
51              //设置日期监听器
52              OnDateSetListener dateListener = new OnDateSetListener()
53              {
54                  @Override
55                  public void onDateSet(DatePicker view, int year,
56                                       int monthOfYear, int dayOfMonth)
57                  {
58                      m_year = year;
59                      m_month = monthOfYear;
60                      m_day = dayOfMonth;
61                  }
62              };
```

```
63              //创建日期对话框对象
64              DatePickerDialog date = new DatePickerDialog(MainActivity.this,
65                       dateListener, m_year, m_month, m_day);
66              date.setTitle("日期对话框");
67              date.show();
68          }
69          else  if(v == btn3)
70          {    //设置时间监听器
71              OnTimeSetListener timeListener = new OnTimeSetListener()
72              {
73                  @Override
74                  public void onTimeSet(TimePicker view, int hourOfDay, int minute)
75                  {
76                      m_hour = hourOfDay;
77                      m_minute = minute;
78                  }
79              };
80              TimePickerDialog d = new TimePickerDialog(MainActivity.this,
81                       timeListener, m_hour, m_minute, true);
82              d.setTitle("时间对话框");
83              d.show();
84          }
85      }
86  }
87 }
```

程序运行结果如图 3.8 所示。

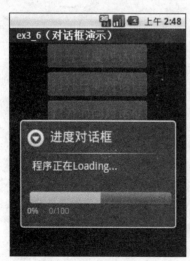

(a) 打开进度对话框　　　　　　(b) 打开日期对话框

图 3.8　对话框示例

习 题 3

1．设计一个具有两个页面的程序，第 1 个页面显示一张封面的图片，第 2 个页面显示"欢迎进入本系统"，这两个页面之间能相互切换。

2．设计一个具有 3 个选项的菜单程序，当单击每个选项时分别跳转到 3 个不同的页面。

3．设计一个具有计算器功能的对话框程序。

第 4 章　图形与多媒体处理

4.1　绘制几何图形

4.1.1　几何图形绘制类

在 Android 系统中绘制几何图形需要用到一些绘图工具，这些绘图工具都在 android.graphics 包中。下面介绍这些绘图工具类的常用方法和属性。

1. 画布类 Canvas

画布类 Canvas 是 Android 绘制几何图形的主要工具，其常用方法如表 4-1 所示。

表 4-1　画布类 Canvas 的常用方法

方法	功能
Canvas()	创建一个空的画布，可以使用 setBitmap() 方法设置绘制具体的画布
Canvas(Bitmap bitmap)	以 bitmap 对象创建一个画布，将内容都绘制在 bitmap 上，bitmap 不得为 null
drawColor()	设置 Canvas 的背景颜色
setBitmap()	设置具体画布
clipRect()	设置显示区域，即设置裁剪区
rotate()	旋转画布
skew()	设置偏移量
drawLine(float x1, float y1, float x2, float y2)	从点（x_1, y_1）到点（x_2, y_2）的直线
drawCircle(float x, float y, float radius, Paint paint)	以（x, y）为圆心、radius 为半径画圆
drawRect(float x1, float y1, float x2, float y2, Paint paint)	从左上角（x_1, y_1）到右下角（x_2, y_2）的矩形
drawText(String text, float x, float y ,Paint paint)	写文字
drawBitmap (Bitmap bitmap, float x, float y, Paint paint)	以左上角坐标（x,y）为顶点绘制 Bitmap 图像
drawPath(Path path, Paint paint)	从一点到另一点的连接路径线段

2. 画笔类 Paint

画笔类 Paint 用来描述所绘制图形的颜色和风格，例如线条宽度、颜色等信息，其常用方法如表 4-2 所示。

3. 点到点的连线路径 Path

在绘制由一些线段组成的图形（例如三角形、四边形等）时需要用 Path 类来描述线段路径。其常用方法如表 4-3 所示。

表 4-2　画笔类 Paint 的常用方法

方法	功能
Paint()	构造方法，创建一个辅助画笔对象
setColor(int color)	设置颜色
setStrokeWidth(float width)	设置画笔宽度
setTextSize(float textSize)	设置文字尺寸
setAlpha(int a)	设置透明度 Alpha 值
setAntiAlias(boolean b)	除去边缘锯齿，取 true 值
paint.setStyle(Paint.Style style)	设置图形为空心（Paint.Style.STROKE）或实心（Paint.Style.FILL）

表 4-3　连线路径 Path 的常用方法

方法	功能
lineTo(float x, float y)	从当前点到指定点画连线
moveTo(float x, float y)	移动到指定点
close()	关闭绘制连线路径

4.1.2　几何图形的绘制过程

在 Android 中绘制几何图形的一般过程如下：
（1）创建一个 View 的子类，并重写 View 类的 onDraw()方法。
（2）在 View 的子类视图中使用画布对象 Canvas 绘制各种图形。
（3）使用 invalidate()方法刷新画面。

【例 4-1】绘制几何图形示例。

本例继承自 Android.view.View 的 TestView 类，重写 View 类的 onDraw()方法，在 onDraw()方法中运用 Paint 对象（绘笔）的不同设置值在 Cavas（画布）上绘制图形，分别绘制了矩形、圆形、三角形和文字。

其源程序如下：

```
1   package com.ex4_1;
2   import android.app.Activity;
3   import android.content.Context;
4   import android.graphics.Canvas;
5   import android.graphics.Color;
6   import android.graphics.Paint;
7   import android.graphics.Path;
8   import android.os.Bundle;
9   import android.view.View;
10  public class MainActivity extends Activity
11  {
12      @Override
13      public void onCreate(Bundle savedInstanceState)
14      {
15          super.onCreate(savedInstanceState);
16          TestView tView=new TestView(this);
17          setContentView(tView);
```

```
18    }
19    private class TestView extends View
20    {
21      public TestView(Context context)
22      {
23          super(context);
24       }
25      /*重写onDraw()*/
26      @Override
27      protected void onDraw(Canvas canvas)
28      {
29       /*设置背景为青色*/
30       canvas.drawColor(Color.CYAN);
31       Paint paint=new Paint();
32       /*设置画笔宽度*/
33       paint.setStrokeWidth(3);
34       /*设置画空心图形*/
35       paint.setStyle(Paint.Style.STROKE);
36       /*去锯齿*/
37       paint.setAntiAlias(true);
38       /*画空心矩形（正方形）*/
39       canvas.drawRect(10,10,70,70,paint);
40       /*设置画实心图形*/
41       paint.setStyle(Paint.Style.FILL);
42       /*画实心矩形（正方形）*/
43       canvas.drawRect(100,10,170,70,paint);
44       /*设置画笔颜色为蓝色*/
45       paint.setColor(Color.BLUE);
46       /*画圆心为(100，120)、半径为30的实心圆*/
47       canvas.drawCircle(100,120,30,paint);
48       /*在实心圆上画一个小白点*/
49       paint.setColor(Color.WHITE);
50       canvas.drawCircle(91,111,6,paint);
51       /*设置画笔颜色为红色*/
52       paint.setColor(Color.RED);
53       /*画三角形*/
54       Path path=new Path();
55       path.moveTo(100, 170);
56       path.lineTo(70, 230);
57       path.lineTo(130,230);
58       path.close();
59       canvas.drawPath(path,paint);
60       /*文字*/
61       paint.setTextSize(28);
62       paint.setColor(Color.BLUE);
```

```
63            canvas.drawText(getResources().getString(R.string. hello_world),
64                    30,270,paint);
65        }
66    }
67 }
```

程序的运行结果如图 4.1 所示。

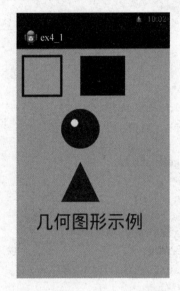

图 4.1　绘制几何图形示例

【例 4-2】绘制一个可以在任意指定位置显示的小球。

设计思想：Android 系统应用程序的设计采用 MVC 模式，即把应用程序分为表现层（View）、控制层（Control）、业务模型层（Model）。在本示例中，按照这种模式图形界面布局为表现层，Activity 控制程序为控制层，实现几何作图的绘制过程属于业务模型层。在业务模型层将圆心坐标设为 (x, y)，则圆的位置随控制层任意输入的坐标值而改变。

其源程序如下：

（1）表现层的图形界面布局程序 activity_main.xml。

```
1  <LinearLayout xmlns:android="http://schemas.android.com/apk/res/ android"
2    android:layout_width="fill_parent"
3    android:layout_height="fill_parent"
4    android:orientation="vertical">
5    <LinearLayout
6      android:layout_width="wrap_content"
7      android:layout_height="wrap_content">
8      <TextView
9         android:id="@+id/ textView1 "
10        android:layout_width="wrap_content "
11        android:layout_height="wrap_content"
```

```
12        android:text="输入位置: "
13        />
14    <EditText
15        android:id="@+id/ editText1 "
16        android:layout_width="120dp"
17        android:layout_height="wrap_content"
18        />
19    <Button
20        android:id="@+id/button1"
21        android:layout_width="wrap_content "
22        android:layout_height="wrap_content"
23        android:text="确定"
24        />
25  </LinearLayout>
26  <com.example.ex4_2.TestView
27       android:id="@+id/testView1 "
28       android:layout_width="match_parent "
29       android:layout_height="match_parent "
30       />
31  </LinearLayout>
```

在界面布局中设置绘制图形的视图组件，在导入自定义组件时要带包名

（2）控制层的主控程序 MainActivity.java。

```
1   package com.example.ex4_2;
2   import android.os.Bundle;
3   import android.view.View;
4   import android.view.View.OnClickListener;
5   import android.widget.Button;
6   import android.widget.EditText;
7   import android.app.Activity;
8   public class MainActivity extends Activity
9   {
10    int x1=150,y1=50;
11    TestView testView;
12    Button btn;
13    EditText edit_y;
14    @Override
15    public void onCreate(Bundle savedInstanceState)
16    {
17        super.onCreate(savedInstanceState);
18        setContentView(R.layout.activity_main);
19        testView=(TestView)findViewById(R.id.testView1);
20        testView.setXY(x1, y1);
21        btn=(Button)findViewById(R.id.button1);
22        edit_y=(EditText)findViewById(R.id.editText1);
23        btn.setOnClickListener(new mClick());
```

设置表现层图形的坐标位置

```
24    }
25    class mClick implements OnClickListener
26    {
27      @Override
28      public void onClick(View arg0)
29      {
30        y1 = Integer.parseInt(edit_y.getText().toString());   ← 字符串转换为整型
31        testView.setXY(x1, y1);   ┐
32        testView.invalidate();    ┘ 在新坐标位置绘制图形,刷新视图
33      }
34    }
35 }
```

对于程序第 30 行:

y1 = Integer.parseInt(edit_y.getText().toString());

方法 Integer.parseInt(String)为将字符串 String 转换为整型数据。
(3) 业务逻辑层的绘制小球程序 TestView.java。

```
1   package com.example.ex4_2;
2   import android.util.AttributeSet;
3   import android.view.View;
4   import android.content.Context;
5   import android.graphics.Canvas;
6   import android.graphics.Color;
7   import android.graphics.Paint;
8
9   public class TestView extends View      ← 继承于 View 的绘制图形类
10  {
11    int x, y;
12    public TestView(Context context, AttributeSet attrs)   ← 在 XML 文件中使用自定义组件时必须使用 AttributeSet 接口对象做参数
13    {
14      super(context, attrs);
15    }
16    void setXY(int _x, int _y)   ← 传递由控制层设置的坐标值
17    {
18      x = _x;
19      y = _y;
20    }
21    @Override
22    protected void onDraw(Canvas canvas)
23    {
24      super.onDraw(canvas);
25      /*设置背景为青色*/
26      canvas.drawColor(Color.CYAN);
```

```
27      Paint paint=new Paint();
28      /*去锯齿*/
29      paint.setAntiAlias(true);
        /*设置paint的颜色*/
30
31      paint.setColor(Color.BLACK);
32      /*画一个实心圆*/
33      canvas.drawCircle(x, y, 15, paint);
34      /*画实心圆上的小白点*/
35      paint.setColor(Color.WHITE);
36      canvas.drawCircle(x-6, y-6, 3, paint);
37    }
38  }
```

程序的运行结果如图 4.2 所示。

图 4.2 在任意指定位置显示小球

4.1.3 自定义组件

在 Android 中可以通过 View 类的子类自定义组件，然后添加到布局界面中。下面通过示例详细说明设计方法。

【例 4-3】 自定义一个组件，再通过布局界面显示出来。
主要设计步骤如下：
（1）编写 View 的子类 TestView。
（2）把 TestView 添加到布局界面中。

(3) 在主程序 MainActivity.java 中建立 TestView 对象与布局文件的关联。

具体操作如下:

(1) 新建 Java 文件,编写 View 的子类 TestView。

```
1   package com.example.ex4_3;
2   import android.content.Context;
3   import android.graphics.Canvas;
4   import android.graphics.Color;
5   import android.graphics.Paint;
6   import android.util.AttributeSet;
7   import android.view.View;
8   class TestView extends View
9   {
10  public TestView(Context context, AttributeSet attrs)
11  {   super(context, attrs);   }
12  /* 重写View的抽象方法 onDraw() */
13  protected void onDraw(Canvas canvas)
14  {
15      canvas.drawColor(Color.CYAN);   //设置组件的背景颜色为青色
16      Paint paint=new Paint();        //定义画笔
17      paint.setStyle(Paint.Style.FILL); //设置画实心图形
18      paint.setAntiAlias(true);       //去锯齿
19      paint.setColor(Color.BLUE);  /*设置画笔颜色为蓝色*/
20      canvas.drawCircle(100,120,30,paint);  /*画圆心为(100,120)、半径为30
        的实心圆*/
21      paint.setColor(Color.WHITE);  /*在实心圆上画一个小白点*/
22      canvas.drawCircle(91,111,6,paint);
23  }
24  }
```

(2) 在表现层布局文件 activity_main.xml 中添加所设计的组件类 TestView。

```
1   <?xml version="1.0" encoding="utf-8"?>
2   <RelativeLayout xmlns:android="http://schemas.android.com/apk/res/android"
3       xmlns:tools="http://schemas.android.com/tools"
4       android:layout_width="match_parent"
5       android:layout_height="match_parent"
6       android:paddingBottom="@dimen/activity_vertical_margin"
7       android:paddingLeft="@dimen/activity_horizontal_margin"
8       android:paddingRight="@dimen/activity_horizontal_margin"
9       android:paddingTop="@dimen/activity_vertical_margin"
10      tools:context="com.example.ex4_3.MainActivity">
11  <com.example.ex4_3.TestView
12          android:id="@+id/testview1"
13          android:layout_width="wrap_content"
```

```
14            android:layout_height="wrap_content"
15        />
16  </RelativeLayout>
```

这时可以在编辑器中预览所自定义的组件,如图 4.3 所示。

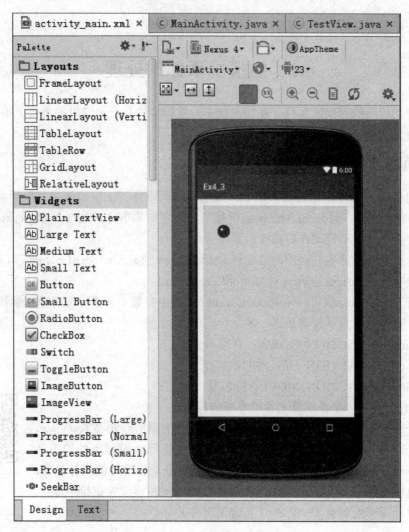

图 4.3　在编辑器中预览自定义的组件

(3) 在主程序 MainActivity.java 中建立 TestView 对象与布局文件的关联。

```
1   package com.example.ex4_3;
2   import android.support.v7.app.AppCompatActivity;
3   import android.os.Bundle;
4   public class MainActivity extends Activity {
5       TestView tView = null;
6       @Override
7       public void onCreate(Bundle savedInstanceState) {
```

```
8        super.onCreate(savedInstanceState);
9        tView =(TestView)findViewById(R.id.testview1);
10       setContentView(R.layout.activity_main);
11   }
12 }
```

运行程序，其运行结果与编辑器中的预览结果一致。

4.2 触摸屏事件的处理

智能移动设备的触摸屏事件（在模拟器中为鼠标事件）分为简单的触摸屏事件和手势识别。下面分别介绍这些事件的处理方法。

4.2.1 简单的触摸屏事件

简单的触摸屏事件是指触摸屏按下、抬起和移动事件（在模拟器中为鼠标事件）。在 Android 系统中通过 OnTouchListener 监听接口来处理屏幕事件，当在 View 的范围内进行触摸按下、抬起或滑动等动作时都会触发该事件。

在设计简单触摸屏事件程序时要实现 android.view.View.OnTouchListener 接口，并重写该接口的监听方法 onTouch(View v, MotionEvent event)。

在监听方法 onTouch(View v, MotionEvent event)中参数 v 为事件源对象，参数 event 为事件对象，事件对象为下列常数之一。

- MotionEvent.ACTION_DOWN：在屏幕上点击。
- MotionEvent.ACTION_UP：抬起。
- MotionEvent.ACTION_MOVE：在屏幕上滑动。

【例 4-4】 设计一个在屏幕上移动小球的程序。

设计一个自定义组件继承于 Android.view.View 的图形绘制类 TestView，在该视图组件中绘制一个小球；再设计一个实现监听触摸屏事件的方法 onTouch(View v, MotionEvent event)，该方法监听并获取触摸屏的坐标位置，并把坐标值传递给图形绘制类 TestView，由 TestView 在该位置重绘小球。

（1）设计图形绘制类 TestView（自定义组件）。

```
1  package com.example.ex4_4;
2  import   (略)
3  class TestView extends View
4  {
5      int x=150,y=50;          ← 定义小球的初始坐标
6      public TestView(Context context, AttributeSet attrs)
7      { super(context, attrs);  }
8      void getXY(int _x, int _y)
9      {                         ← 由触摸屏事件传递小球坐标位置
10         x = _x;
11         y = _y;
```

```
12      }
13     @Override
14     protected void onDraw(Canvas canvas)
15     {
16         super.onDraw(canvas);
17         canvas.drawColor(Color.CYAN); /*设置背景为青色*/
18         Paint paint=new Paint();
19         paint.setAntiAlias(true); /*去锯齿*/
20         paint.setColor(Color.BLACK); /*设置paint的颜色*/
21         canvas.drawCircle(x, y, 30, paint); /*画一个实心圆*/
22         paint.setColor(Color.WHITE);/*画实心圆上的小白点*/
23         canvas.drawCircle(x-9, y-9, 6, paint);
24     }
```

(2) 把自定义组件添加到布局文件 activity_main.xml 中。

```
1  <?xml version="1.0" encoding="utf-8"?>
2  <RelativeLayout xmlns:android="http://schemas.android.com/apk/res/android"
3    xmlns:tools="http://schemas.android.com/tools"
4    android:layout_width="match_parent"
5    android:layout_height="match_parent"
6    android:paddingBottom="@dimen/activity_vertical_margin"
7    android:paddingLeft="@dimen/activity_horizontal_margin"
8    android:paddingRight="@dimen/activity_horizontal_margin"
9    android:paddingTop="@dimen/activity_vertical_margin"
10   tools:context="com.example.ex4_4.MainActivity">
11     <com.example.ex4_4.TestView
12         android:id="@+id/testview1"
13         android:layout_width="wrap_content"
14         android:layout_height="wrap_content"
15     />
16 </RelativeLayout>
```

← 添加自定义组件

(3) 设计主控文件 MainActivity.java。

```
1  package com.example.ex4_4;
2  import android.support.v7.app.AppCompatActivity;
3  import android.os.Bundle;
4  public class MainActivity extends Activity {
5      TestView tView = null;
6      @Override
7      public void onCreate(Bundle savedInstanceState) {
8          super.onCreate(savedInstanceState);
9          tView =(TestView)findViewById(R.id.testview1);
10         tView.setOnTouchListener(new mOnTouch());
```

```
11          setContentView(R.layout.activity_main);
12      }
13  class mOnTouch implementsView.OnTouchListener {
14      @Override
15      public boolean onTouch(View v, MotionEvent event) {
16       int x1,y1;
17       x1 = (int) event.getX();         ← 获取坐标位置
18       y1 = (int) event.getY();
19       if(event.getAction() == MotionEvent.ACTION_DOWN)
20       {   tView.get_xy(x1, y1);        ← 按新坐标绘图    ← 在屏幕上点击
21           tView.invalidate();
22           return true;
23       }
24       else if(event.getAction() == MotionEvent.ACTION_MOVE)
25       {   tView.get_xy(x1, y1);        ← 按新坐标绘图    ← 在屏幕上滑动（拖动）
26           tView.invalidate();
27           return true;
28       }
29       return   tView.onTouchEvent(event);
30      }
31  }
32 }
```

程序的运行结果如图 4.4 所示，用手指（或鼠标）在屏幕上滑动时小球将随手指移动，当点击屏幕时小球将被移动到所点击位置。

图 4.4 用手指（或鼠标）在屏幕上滑动时小球随之移动

【例 4-5】 设计一个能在图片上涂鸦的程序。
（1）设计布局文件。

```
1  <?xml version="1.0" encoding="utf-8"?>
2  <LinearLayout xmlns:android="http://schemas.android.com/apk/
   res/android"
```

```xml
3      android:layout_width="fill_parent"
4      android:layout_height="fill_parent"
5      android:orientation="vertical" >
6      <com.ex4_5.HandWrite
7         android:id="@+id/handwriteview"
8         android:layout_width="fill_parent"
9         android:layout_height="380dp" />
10     <LinearLayout
11        android:layout_width="fill_parent"
12        android:layout_height="fill_parent"
13        android:orientation="horizontal"
14        android:gravity="center_horizontal" >
15        <Button
16           android:id="@+id/clear"
17           android:layout_width="200dp"
18           android:layout_height="wrap_content"
19           android:text="清屏" />
20     </LinearLayout>
21  </LinearLayout>
```

注释: 导入自定义 View,注意要带包

(2) 设计主控文件 MainActivity.java。

```java
1   package com.ex4_5;
2   import android.app.Activity;
3   import android.os.Bundle;
4   import android.view.View;
5   import android.view.View.OnClickListener;
6   import android.widget.Button;
7   public class MainActivity extends Activity
8   {
9     private HandWrite handWrite = null;
10    private Button clear = null;
11    @Override
12    public void onCreate(Bundle savedInstanceState)
13    {
14      super.onCreate(savedInstanceState);
15      setContentView(R.layout.main);
16      handWrite = (HandWrite)findViewById(R.id.handwriteview);
17      clear = (Button)findViewById(R.id.clear);
18      clear.setOnClickListener(new mClick());
19    }
20    private class mClick implements OnClickListener
21    {
22      public void onClick(View v)
23      {
24        handWrite.clear();
```

注释: 关联 View 组件; 清屏

（3）记录在屏幕上滑动的轨迹，实现在图片上涂鸦的功能。

```
1    package com.ex4_5;
2    import android.content.Context;
3    import android.graphics.*;
4    import android.graphics.Paint.Style;
5    import android.util.AttributeSet;
6    import android.view.MotionEvent;
7    import android.view.View;
8    public class HandWrite extends View      ◄── 自定义View组件HandWrite
9    {
10       Paint paint = null;                  //定义画笔
11       Bitmap originalBitmap = null;        //存放原始图像
12       Bitmap new1_Bitmap = null;           //存放从原始图像复制的位图图像
13       Bitmap new2_Bitmap = null;           //存放处理后的图像
14       float startX = 0,startY = 0;         //画线的起点坐标
15       float clickX = 0,clickY = 0;         //画线的终点坐标
16       boolean isMove = true;               //设置是否画线的标记
17       boolean isClear = false;             //设置是否清除涂鸦的标记
18       int color = Color.GREEN;             //设置画笔的颜色（绿色）
19       float strokeWidth = 2.0f;            //设置画笔的宽度
20       public HandWrite(Context context, AttributeSet attrs)
21       {
22          super(context, attrs);
23          originalBitmap = BitmapFactory    ◄── 从资源中获取原始图像
24               .decodeResource(getResources(), R.drawable.cy)
                 .copy(Bitmap.Config.ARGB_8888, true);
25          new1_Bitmap = Bitmap.createBitmap(originalBitmap);  ◄── 建立原始图像的位图
26       }
27       public void clear(){
28          isClear = true;
29          new2_Bitmap = Bitmap.createBitmap(originalBitmap);   ── 清除涂鸦
30          invalidate();
31       }
32       public void setstyle(float strokeWidth){
33          this.strokeWidth = strokeWidth;
34       }
35       @Override
36       protected void onDraw(Canvas canvas)
37       {
38          super.onDraw(canvas);                                ── 显示绘图
39          canvas.drawBitmap(HandWriting(new1_Bitmap), 0, 0,null);
40       }
```

```java
41  public Bitmap HandWriting(Bitmap o_Bitmap)    // 记录绘制图形
42  {
43      Canvas canvas = null;    // 定义画布
44      if(isClear)
45      {
46          canvas = new Canvas(new2_Bitmap);    // 创建绘制新图形的画布
47      }
48      else{
49          canvas = new Canvas(o_Bitmap);    // 创建绘制原图形的画布
50      }
51      paint = new Paint();
52      paint.setStyle(Style.STROKE);
53      paint.setAntiAlias(true);                // 定义画笔
54      paint.setColor(color);
55      paint.setStrokeWidth(strokeWidth);
56     if(isMove)
57     {
58      canvas.drawLine(startX, startY, clickX, clickY, paint);   // 在画布上画线条
59     }
60     startX = clickX;
61     startY = clickY;
62     if(isClear)
63     {
64         return new2_Bitmap;    // 返回新绘制的图像
65     }
66     return o_Bitmap;    // 若清屏,则返回原图像
67  }
68  @Override
69  public boolean onTouchEvent(MotionEvent event)    // 定义触摸屏事件
70  {
71     clickX = event.getX();        // 获取触摸坐标位置
72     clickY = event.getY();
73     if(event.getAction() == MotionEvent.ACTION_DOWN)    // 按下屏幕时无绘图
74     {
75        isMove = false;
76        invalidate();
77        return true;
78     }
79     else if(event.getAction() == MotionEvent.ACTION_MOVE)
80     {
81        isMove = true;              // 记录在屏幕上滑动的轨迹
82        invalidate();
83        return true;
84     }
85     return super.onTouchEvent(event);
```

```
86    }
87  }
```

程序的运行结果如图 4.5 所示,当手指(或鼠标)在屏幕上滑动时记录下滑动的轨迹,在图片上涂鸦,点击"清屏"按钮则清除图片上的痕迹。

(原图)　　　　　　　　　　(涂鸦)

图 4.5　在图片上涂鸦

4.2.2　手势识别

所谓手势识别,就是识别手指(或鼠标)在屏幕上滑动时的轨迹。在 Android 系统中 android.gesture 是用于创建、识别和保存触摸屏手势功能的包。android.gesture 包的主要类及接口如表 4-4 所示。

表 4-4　android.gesture 包的主要类及接口

类及接口	功能
Gesture	触摸屏的手势类
GestureLibraries	手势库
GestureLibrary	手势库
GestureOverlayView	可输入手势的视图
Prediction	手势的预显示类
GestureStroke	记录触摸屏上手势动作的开始与结束类
OnGestureListener	手势动作的监听接口
OnGesturePerformedListener	可输入手势视图 GestureOverlayView 的监听接口

在实现 OnGesturePerformedListener 接口时需要覆盖以下方法:

```
onGesturePerformed(GestureOverlayView overlay, Gesture gesture)
```

【例 4-6】 设计一个手写字体识别程序。

如果要编写一个手写字体识别程序,必须先建立一个存放手写字体的数据库。在手机模拟器中已经预装了创建手写字体数据库的应用程序 Gestures Builder,其图标如图 4.6 所示。

图 4.6 建立手写字体数据库 Gestures Builder 的图标

创建手势库如图 4.7 所示。由手势创建的手写字体将被保存到 sdcard\gestures 中,把文件 gestures 复制到项目 res\raw 下,这样就可以在应用程序里面使用这些手势了。

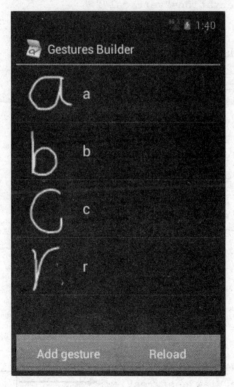

图 4.7 创建手势库

(1) 设计界面布局文件 activity_main.xml。在界面布局文件 activity_main.xml 中设置 android.gesture.GestureOverlayView 组件,其代码如下:

```
1  <?xml version="1.0" encoding="utf-8"?>
2                                            <LinearLayout
```

```
    xmlns:android="http://schemas.android.com/apk/res/android"
3   android:layout_width="fill_parent"
4   android:layout_height="fill_parent"
5   android:orientation="vertical" >
6   <TextView
7     android:id="@+id/textView1"
8     android:layout_width="fill_parent"
9     android:layout_height="wrap_content"
10    android:text="@string/text1"
11    android:textSize="24sp"/>
12    <!-- 绘制手势的GestureOverlayView -->
13  <android.gesture.GestureOverlayView       ← 设置 View 组件，带包名
14    android:id="@+id/gestures1"
15    android:layout_width="fill_parent"
16    android:layout_height="fill_parent"
17    android:gestureStrokeType="multiple"
18    android:eventsInterceptionEnabled="false"
19    android:orientation="vertical"/>
20  </LinearLayout>
```

（2）设计控制程序 MainActivity.java。

```
1  package com.ex4_6;
2  import java.util.ArrayList;
3  import android.app.Activity;
4  import android.gesture.Gesture;
5  import android.gesture.GestureLibraries;
6  import android.gesture.GestureLibrary;
7  import android.gesture.GestureOverlayView;
8  import android.gesture.Prediction;
9  import android.gesture.GestureOverlayView.OnGesturePerformedListener;
10 import android.os.Bundle;
11 import android.widget.TextView;
12 import android.widget.Toast;
13 public class MainActivity extends Activity
14 implements OnGesturePerformedListener
15 {
16   GestureLibrary mLibrary;              ← 定义手势库对象
17   GestureOverlayView gesturesView;      ← 定义手势视图对象做画板之用
18   TextView txt;
19   @Override
20   public void onCreate(Bundle savedInstanceState)
21   {
22     super.onCreate(savedInstanceState);
23     setContentView(R.layout.main);
```

```
24    gesturesView=(GestureOverlayView)findViewById(R.id.gestures);
25    gesturesView.addOnGesturePerformedListener(this);    ← 注册手势识别的监听器
26    txt = (TextView)findViewById(R.id.textView1);
27    mLibrary = GestureLibraries.fromRawResource(this,
28                     R.raw.gestures);
29    if(!mLibrary.load())                                 ← 加载手势库
30    {
31      finish();
32    }
33  }
34  /* 根据在GestureOverlayView上画的手势来识别是否匹配手势库里的手势 */
35  @Override
36  public void onGesturePerformed(GestureOverlayView overlay, Gesture gesture)
37  {
38    ArrayList predictions=mLibrary.recognize(gesture);   ← 从手势库中获取手势数据
39    if(predictions.size()>0)
40    {
41      Prediction prediction = (Prediction)predictions.get(0);
42      if(prediction.score > 1.0)    ← 检索到匹配的手势
43      {
44        Toast.makeText(this,prediction.name,Toast.LENGTH_SHORT). show();
45        txt.append(prediction.name);
46      }
47    }
48  }
49  }
```

程序的运行结果如图 4.8 所示。

图 4.8 手写字体识别

4.3 音频播放

4.3.1 多媒体处理包

Android 系统提供了针对常见多媒体格式的 API，可以非常方便地操作图片、音频、视频等多媒体文件，也可以操纵 Android 终端的录音、摄像设备。这些多媒体处理 API 均位于 android.media 包中。android.media 包中的主要类如表 4-5 所示。

表 4-5 android.media 包中的主要类

类名或接口名	说明
MediaPlayer	支持流媒体，用于播放音频和视频
MediaRecorder	用于录制音频和视频
Ringtone	用于播放可用作铃声和提示音的短声音片段
AudioManager	负责控制音量
AudioRecord	用于记录从音频输入设备产生的数据
JetPlayer	用于存储 JET 内容的回放和控制
RingtoneManager	用于访问响铃、通知和其他类型的声音
Ringtone	快速播放响铃、通知或其他相同类型的声音
SoundPool	用于管理和播放应用程序的音频资源

4.3.2 多媒体处理播放器 MediaPlayer

1. MediaPlayer 类的常用方法

MediaPlayer 是 Android 系统多媒体 android.media 包中的类，MediaPlayer 类主要用于控制音频文件、视频文件或流媒体的播放。MediaPlayer 类的常用方法如表 4-6 所示。

表 4-6 MediaPlayer 类的常用方法

方法	说明
create()	创建多媒体播放器
getCurrentPosition()	获得当前播放位置
getDuration()	获得播放文件的时间
getVideoHeight()	播放视频的高度
getVideoWidth()	播放视频的宽度
isLooping()	是否循环播放
isPlaying()	是否正在播放
Pause()	暂停
Prepare()	准备播放文件，进行同步处理
prepareAsync()	准备播放文件，进行异步处理
release()	释放 MediaPlayer 对象
reset()	重置 MediaPlayer 对象
seekTo()	指定播放文件的播放位置
setDataSource()	设置多媒体数据来源
setVolume()	设置音量
setOnCompletionListener()	监听播放文件播放完毕
start()	开始播放
stop()	停止播放

2. MediaPlayer 对象的生命周期

通常把一个对象从创建、使用直到释放的过程称为该对象的生命周期，把 MediaPlayer 对象的创建、初始化、同步处理、开始播放、播放结束的运行过程称为 MediaPlayer 的生命周期。MediaPlayer 对象的生命周期如图 4.9 所示。

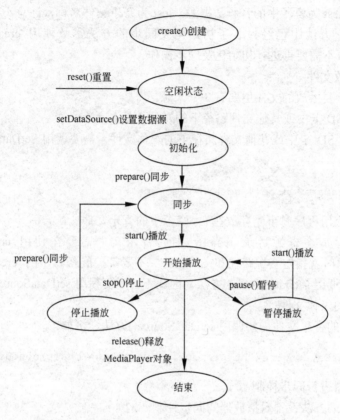

图 4.9　MediaPlayer 对象的生命周期

从图 4.9 可以看出，当一个 MediaPlayer 对象刚创建或者调用了 reset() 方法后，它处于 Idle（空闲）状态；当调用了 release() 方法后，它处于释放（结束）状态。这两种状态之间是 MediaPlayer 对象的生命周期。一个 MediaPlayer 对象在处于空闲状态时是不能进行播放操作的，还必须经过初始化、同步阶段才能进行播放操作。

4.3.3 播放音频文件

通过媒体处理器 MediaPlayer 提供的方法不仅可以播放存放在 SD 卡上的音乐文件，而且还能播放资源中的音乐文件。这二者在设计方法上稍有不同。

下面按上述两种情况来说明应用 MediaPlayer 对象播放音频文件的步骤。

1. 构建 MediaPlayer 对象

1）使用 new 方式创建 MediaPlayer 对象

播放 SD 卡上的音乐文件需要使用 new 方式创建 MediaPlayer 对象：

```
MediaPlayer mplayer = new MediaPlayer();
```

2）使用 create 方法创建 MediaPlayer 对象

播放资源中的音乐需要使用 create 方法创建 MediaPlayer 对象，例如：

```
MediaPlayer mplayer = MediaPlayer.create(this, R.raw.test);
```

其中 R.raw.test 为资源中的音频数据源，test 为音乐文件名称，注意不要带扩展名。

由于 create 方法中已经封装了初始化及同步的方法，故使用 create 方法创建的 MediaPlayer 对象不需要再进行初始化及同步操作。

2. 设置播放文件

MediaPlayer 要播放的文件主要有 3 个来源。

1）存储在 SD 卡中或其他文件路径下的媒体文件

对于存储在 SD 卡中或其他文件路径下的媒体文件，需要调用 setDataSource()方法，例如：

```
mplayer.setDataSource("/sdcard/test.mp3");
```

2）在编写应用程序时事先存放在 res 资源中的音乐文件

播放事先存放在资源目录 res\raw 中的音乐文件需要在使用 create()方法创建 MediaPlayer 对象时就指定资源路径和文件名称（不要带扩展名）。由于 create()方法的源代码中已经封装了调用 setDataSource()方法，因此不必重复使用 setDataSource()方法。

3）网络上的媒体文件

播放网络上的音乐文件需要调用 setDataSource()方法，例如：

```
mplayer.setDataSource("http://www.citynorth.cn/music/confucius.mp3");
```

3. 对播放器进行同步控制

使用 prepare()方法设置对播放器的同步控制，例如：

```
mplayer.prepare();
```

如果 MediaPlayer 对象是由 create 方法创建的，由于 create()方法的源代码中已经封装了调用 prepare()方法，因此可省略此步骤。

4. 播放音频文件

start()是真正启动音频文件播放的方法，例如：

```
mplayer.start();
```

如要暂停播放或停止播放，则调用 pause()和 stop()方法。

5. 释放占用资源

在音频文件播放结束时，应该调用 release()释放播放器占用的系统资源。

如果要重新播放音频文件，则需要调用 reset()返回到空闲状态，再从第 2 步开始重复其他各步骤。

【例 4-7】 设计一个音乐播放器。

在本示例中将分别播放存放在项目资源中的音乐文件和 SD 卡中的音频文件，因此需要事先将准备好的音频文件保存在指定路径下。

（1）将测试的音频文件 mtest1.mp3 复制到新建项目的 res\raw 目录下。

（2）将音频文件 mtest2.mp3 复制到 SD 卡中（在模拟器中使用 SD 卡，可以在 Eclipse 集成环境中选择 DDMS 调试工具，单击"向设备导入文件"按钮，将音频文件复制到模拟器的 mntsdcard/Music 目录下，如图 4.10 所示）。

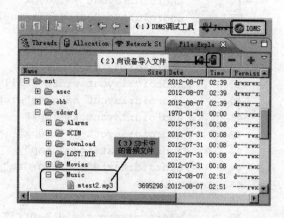

图 4.10　将音频文件存放到模拟器的 SD 卡中

（1）设计布局文件 main.xml。在用户界面布局中设置 3 个带图标的按钮，分别表示播放、暂停、停止，再设置两个选项按钮，以选择播放的文件。布局文件的代码如下：

```xml
1  <?xml version="1.0" encoding="utf-8"?>
2  <AbsoluteLayout
3      xmlns:android="http://schemas.android.com/apk/res/android"
4      android:orientation="vertical"
5      android:layout_width="fill_parent"
6      android:layout_height="fill_parent">
7      <TextView
8          android:id="@+id/text1"
9          android:layout_width="fill_parent"
10         android:layout_height="wrap_content"
11         android:layout_x="5px"
12         android:layout_y="10px"
13         android:text="@string/hello"
14         android:textSize="20sp"/>
15     <ImageButton                     ← 带图标的按钮，需要设置图标的路径
16         android:id="@+id/Stop"
17         android:layout_height="wrap_content"
18         android:layout_width="wrap_content"
19         android:layout_x="30px"
20         android:layout_y="100px"
21         android:src="@drawable/music_stop"   />  ← 设置图标的路径和文件名称
22     <ImageButton   ← 带图标的按钮
23         android:id="@+id/Start"
24         android:layout_height="wrap_content"
25         android:layout_width="wrap_content"
26         android:layout_x="90px"
27         android:layout_y="100px"
28         android:src="@drawable/music_play"   />  ← 设置图标的路径和文件名称
29     <ImageButton   ← 带图标的按钮
30         android:id="@+id/Pause"
31         android:layout_height="wrap_content"
32         android:layout_width="wrap_content"
```

```
33        android:layout_x="150px"
34        android:layout_y="100px"
35        android:src="@drawable/music_pause"   />   ← 设置图标的路径和文件名称
36
37    <CheckBox
38        android:id="@+id/check1"
39        android:layout_width="fill_parent"
40        android:layout_height="wrap_content"
41        android:layout_x="10px"
42        android:layout_y="180px"
43        android:textSize="20sp"
44        android:text="@string/one" />
45    <CheckBox
46        android:id="@+id/check2"
47        android:layout_width="fill_parent"
48        android:layout_height="wrap_content"
49        android:layout_x="10px"
50        android:layout_y="210px"
51        android:textSize="20sp"
52        android:text="@string/two" />
53 </AbsoluteLayout>
```

(2) 设计控制文件。

```
1  package com.ex4_7;
2  import java.io.IOException;
3  import android.app.Activity;
4  import android.media.MediaPlayer;
5  import android.os.Bundle;
6  import android.util.Log;
7  import android.view.View;
8  import android.view.View.OnClickListener;
9  import android.widget.CheckBox;
10 import android.widget.ImageButton;
11 import android.widget.TextView;
12 public class MainActivity extends Activity
13 {
14     CheckBox ch1,ch2;              ← 选项按钮
15     TextView txt;
16     ImageButton mStopButton, mStartButton, mPauseButton;   ← 播放控制按钮
17     MediaPlayer mMediaPlayer;          ← MediaPlayer 对象
18     String sdcard_file;
19     sdcard_file = new String("/sdcard/music/mtest2.mp3");   ← SD 卡中的文件
20     int res_file = R.raw.mtest1;       ← 设置资源文件 mtest1
21     @Override
22     public void onCreate(Bundle savedInstanceState)
23     {
24         super.onCreate(savedInstanceState);
25         setContentView(R.layout.main);
26         /* 构建MediaPlayer对象 */
```

```
27      mMediaPlayer = new MediaPlayer();
28      ch1=(CheckBox)findViewById(R.id.check1);
29      ch2=(CheckBox)findViewById(R.id.check2);
30      txt = (TextView)findViewById(R.id.text1);
31      mStopButton = (ImageButton) findViewById(R.id.Stop);
32      mStartButton = (ImageButton) findViewById(R.id.Start);
33      mPauseButton = (ImageButton) findViewById(R.id.Pause);
34      setRegist();   //动态获取本地存储卡（SD卡）读写权限
35      mStopButton.setOnClickListener(new mStopClick());
36      mStartButton.setOnClickListener(new mStartClick());
37      mPauseButton.setOnClickListener(new mPauseClick());
38      }
   //播放SD卡或其他路径的音乐文件
39   private void playMusic(String path)       参数 path 为文件路径
40   {
41      try
42      {
          /* 重置MediaPlayer对象，使之处于空闲状态 */
43         mMediaPlayer.reset();
          /* 设置要播放文件的路径 */
44         mMediaPlayer.setDataSource(path);
          /* 准备播放 */
45         mMediaPlayer.prepare();
          /* 开始播放 */
46         mMediaPlayer.start();
47      }catch (IOException e){    }
48   }
   /* 停止按钮事件 */
49   class mStopClick implements OnClickListener
50   {
51      @Override
52      public void onClick(View v)
53      {
          /* 是否正在播放 */
54         if (mMediaPlayer.isPlaying())
55         {
             //重置MediaPlayer到初始状态
56            mMediaPlayer.reset();
57            mMediaPlayer.release();      释放占用的资源
58         }
59      }
60   }
   /* 播放按钮事件 */
61   class mStartClick implements OnClickListener
62   {
63      @Override
64      public void onClick(View v)
65      {
66   String str="";
67         if(ch1.isChecked())      选择播放系统资源中的音乐
68         {
```

组件的初始化

```
69        str = str + "\n" + ch1.getText();        ← 提示信息
70        try {
71          mMediaPlayer = MediaPlayer.create(MainActivity.this, res_file);
72          mMediaPlayer.start();        ← 开始播放系统资源中的音乐
73        } catch (Exception e) {Log.i("ch1", "res err ..."); }
74     }
75     if(ch2.isChecked())        ← 选择播放 SD 卡的音频文件
76     {
77        str=str + "\n" + ch2.getText();        ← 提示信息
78        try{
79          mMediaPlayer = new MediaPlayer();
80          mMediaPlayer.setDataSource(sdcard_file);        ← 设置音乐源
81          playMusic(sdcard_file);        ← 调用第 38 行的播放方法
82        } catch (Exception e){ Log.i("ch2", "sdcard err ..."); }
83     }
84     txt.setText(str);        ← 显示提示信息
85    }
86  }
    /* 暂停按钮事件  */
87  class mPauseClick implements OnClickListener
88  {
89     @Override
90     public void onClick(View v)
91     {
92        if (mMediaPlayer.isPlaying())
93        {
            /* 暂停 */        ← 第 1 次按暂停键
94          mMediaPlayer.pause();
95        }
96        else
97        {
            /* 开始播放 */        ← 重复按暂停键
98          mMediaPlayer.start();
99        }
100    }
101 }
    //动态获取本地存储卡（SD卡）读写权限
102 private void setRegist() {
    /*
       * 动态获取权限
       * Android 6.0 新特性，一些保护权限，如，文件读写除了要在
       *AndroidManifest中声明权限，还要使用如下代码动态获取
       */
103 if (Build.VERSION.SDK_INT >= 23) {//大于23是指Android 6.0以后版本
104 int REQUEST_CODE_CONTACT = 101;
105 final int REQUEST_EXTERNAL_STORAGE = 1;
106     String[] PERMISSIONS_STORAGE = {
107         Manifest.permission.READ_EXTERNAL_STORAGE,
108         Manifest.permission.WRITE_EXTERNAL_STORAGE
```

```
109        };
       //验证是否许可权限
110    for (String str : PERMISSIONS_STORAGE) {
111        if (this.checkSelfPermission(str) !=
112                            PackageManager.PERMISSION_GRANTED){
       //申请权限
113        this.requestPermissions(PERMISSIONS_STORAGE,
114                            REQUEST_CODE_CONTACT);
115        return;
116        }
117    }
118    }
119 }
```

音频播放的程序运行结果如图 4.11 所示。

图 4.11　音频播放示例

4.4　视　频　播　放

在 Android 系统中设计播放视频的应用程序有两种方式，一种方式是应用媒体播放器组件 MediaPlayer 播放视频，另一种方式是应用视频视图组件 VideoView 播放视频。下面分别介绍这两种设计方式。

4.4.1　应用媒体播放器播放视频

使用媒体播放器组件 MediaPlayer 不仅可以播放音频文件，而且可以播放格式为.3gp

的视频文件。与播放音频的不同之处为用于视频播放的播放承载体必须是实现了表面视图处理接口（surfaceHolder）的视图组件，即需要使用 SurfaceView 组件来显示播放的视频图像。

【例 4-8】 应用媒体播放器组件 MediaPlayer 设计一个视频播放器。

（1）事先准备视频文件 sample.3gp，并将其复制到模拟器的 SD 卡的 sdcard\zsm 目录下。

（2）设计布局文件 activity_main.xml。在用户界面布局中设置一个 SurfaceView 组件，用于显示视频图像；再设置一个按钮，点击按钮开始播放视频文件。布局文件的源代码如下：

```xml
1   <?xml version="1.0" encoding="utf-8"?>
2   <LinearLayout xmlns:android="http://schemas.android.com/apk/res/android"
3       android:layout_width="fill_parent"
4       android:layout_height="fill_parent"
5       android:orientation="vertical" >
6       <TextView
7           android:id="@+id/TextView01"
8           android:layout_width="wrap_content"
9           android:layout_height="wrap_content"
10          android:layout_gravity="center_horizontal"
11          android:text="媒体播放器"
12          android:textSize="24sp" />
13      <SurfaceView              ← 用于显示视频图像
14          android:id="@+id/surfaceView1"
15          android:layout_width="240dp"
16          android:layout_height="320dp"
17          android:layout_gravity="center" />
18      <Button
19          android:id="@+id/play1"
20          android:layout_width="80dp"
21          android:layout_height="40dp"
22          android:text="播放"
23          android:textSize="18sp"   />
24  </LinearLayout>
```

（3）设计控制文件 MainActivity.java。

```java
1   package com.ex4_8;
2
3   import android.media.AudioManager;
4   import android.media.MediaPlayer;
5   import android.os.Bundle;
6   import android.util.Log;
7   import android.view.SurfaceHolder;
```

```
8   import android.view.SurfaceView;
9   import android.view.View;
10  import android.view.View.OnClickListener;
11  import android.widget.Button;
12  import android.app.Activity;
13
14  public class MainActivity extends Activity
15  {
16    MediaPlayer mMediaPlayer;
17    SurfaceView mSurfaceView;
18    Button playBtn;
19    String path;
20    SurfaceHolder sh;          ◄── 表面视图处理接口对象
21
22    @Override
23    public void onCreate(Bundle savedInstanceState)
24    {
25       super.onCreate(savedInstanceState);
26       setContentView(R.layout.Activity_main);
27       mSurfaceView = (SurfaceView)findViewById(R.id.surfaceView1);
28       playBtn=(Button)findViewById(R.id.play1);
29       path = "/sdcard/zsm/sample.3gp";      ◄── 设定视频文件路径
30       mMediaPlayer = new MediaPlayer();
31       playBtn.setOnClickListener(new mClick());
32    }
33
34    class mClick implements OnClickListener
35    {
36       @Override
37       public void onClick(View v)
38       {
39       try {
40            mMediaPlayer.reset();
41         //为播放器对象设置用于显示视频内容、代表屏幕描绘的控制器
42            mMediaPlayer.setAudioStreamType(AudioManager.STREAM_MUSIC);
43            mMediaPlayer.setDataSource(path);//设置数据源
44            sh=mSurfaceView.getHolder();   ◄── 创建表面视图处理接口对象
45            mMediaPlayer.setDisplay(sh);
46            mMediaPlayer.prepare();       ◄── MediaPlayer 对象同步
47            mMediaPlayer.start();
48          }catch (Exception e){ Log.i("MediaPlay err", "MediaPlay err");}
49       }
50    }
51  }
```

视频播放程序的运行结果如图4.12所示。

图4.12　应用媒体播放器组件MediaPlayer设计的视频播放器

4.4.2　应用视频视图播放视频

在Android系统中经常使用android.widget包中的视频视图类VideoView播放视频文件。VideoView类可以从不同的来源（例如资源文件或内容提供器）读取图像，计算和维护视频的画面尺寸以使其适用于任何布局管理器，并提供一些诸如缩放、着色之类的显示选项。VideoView类的常用方法如表4-7所示。

表4-7　VideoView类的常用方法

方法	说明
VideoView(Context context)	创建一个默认属性的VideoView实例
boolean canPause()	判断是否能够暂停播放视频
int getBufferPercentage()	获得缓冲区的百分比
int getCurrentPosition()	获得当前的位置
int getDuration()	获得所播放视频的总时间
boolean isPlaying()	判断是否正在播放视频
boolean onTouchEvent (MotionEvent ev)	实现该方法来处理触屏事件
seekTo (int msec)	设置播放位置
setMediaController(MediaController controller)	设置媒体控制器
setOnCompletionListener(MediaPlayer.OnCompletionListenerl)	注册在媒体文件播放完毕时调用的回调函数
setOnPreparedListener(MediaPlayer.OnPreparedListener l)	注册在媒体文件加载完毕可以播放时调用的回调函数
setVideoPath(String path)	设置视频文件的路径名
setVideoURI(Uri uri)	设置视频文件的统一资源标识符
start()	开始播放视频文件
stopPlayback()	停止回放视频文件

【例 4-9】 应用视频视图组件 VideoView 设计一个视频播放器。

(1) 创建用户界面。新建工程 ex4_9,修改 res\layout 中的 activity_main.xml 布局文件,在里面添加一个显示视图 VideoView 和一个按钮 Button。完整的布局文件 activity_main.xml 如下:

```xml
1  <?xml version="1.0" encoding="utf-8"?>
2  <LinearLayout xmlns:android="http://schemas.android.com/apk/res/android"
3    android:layout_width="fill_parent"
4    android:layout_height="fill_parent"
5    android:orientation="vertical" >
6    <TextView
7      android:layout_width="fill_parent"
8      android:layout_height="wrap_content"
9      android:textSize="24sp"
10     android:text="@string/hello" />
11   <VideoView
12     android:id="@+id/video"
13     android:layout_width="320dp"
14     android:layout_height="240dp" />
15   <Button
16     android:id="@+id/playButton"
17     android:layout_width="wrap_content"
18     android:layout_height="wrap_content"
19     android:text="@string/playButton"
20     android:textSize="20sp"   />
21 </LinearLayout>
```

(2) 设计控制程序 MainActivity.java。

```java
1  package com.ex4_9;
2  import android.app.Activity;
3  import android.os.Bundle;
4  import android.view.View;
5  import android.view.View.OnClickListener;
6  import android.widget.Button;
7  import android.widget.MediaController;
8  import android.widget.VideoView;
9  public class MainActivity extends Activity
10 {
11   private VideoView mVideoView;
12   private Button playBtn;
13   MediaController mMediaController;
14   @Override
15   public void onCreate(Bundle savedInstanceState)
16   {
```

```
17        super.onCreate(savedInstanceState);
18        setContentView(R.layout.Activity_main);
19        mVideoView = new VideoView(this);
20        mVideoView = (VideoView)findViewById(R.id.video);
21        mMediaController = new MediaController(this);
22        playBtn = (Button)findViewById(R.id.playButton);
23        playBtn.setOnClickListener(new mClick());
24    }
25    class mClick implements OnClickListener
26    {
27        @Override
28        public void onClick(View v)
29        {
30            String path="/sdcard/test.3gp";
31            mVideoView.setVideoPath(path);
32            mMediaController.setMediaPlayer(mVideoView);    ← 设置媒体控
33            mVideoView.setMediaController(mMediaController);   制器
34            mVideoView.start();
35        }
36 }
37 }
```

视频播放程序的运行结果如图 4.13 所示。

图 4.13 应用视频视图组件 VideoView 设计的视频播放器

4.5 录音与拍照

4.5.1 用于录音、录像的 MediaRecorder 类

应用 android.media 包中的 MediaRecorder 类可以录制音频和视频。下面详细介绍 MediaRecorder 类的使用方法。

1. MediaRecorder 类的常用方法

MediaRecorder 类的常用方法如表 4-8 所示。

表 4-8 MediaRecorder 类的常用方法

方法	说明
MediaRecorder()	创建录制媒体对象
setAudioSource(int audio_source)	设置音频源
setAudioEncoder(int audio_encoder)	设置音频编码格式
setVideoSource(int video_source)	设置视频源
setVideoEncoder(int video_encoder)	设置视频编码格式
setVideoFrameRate(int rate)	设置视频帧速率
setVideoSize(int width, int height)	设置视频录制画面大小
setOutputFormat(int output_format)	设置输出格式
setOutputFile(path)	设置输出文件路径
prepare()	录制准备
start()	开始录制
stop()	停止录制
reset()	重置
release()	释放播放器有关资源

2. MediaRecorder 对象的数据采集源

在使用音频输入设备（麦克风）进行录音时录音接口支持的音频源类型如下。

- DEFAULT：系统音频源。
- MIC：麦克风。

在使用摄像设备进行视频录制时摄像机接口所支持的视频源类型如下。

- CAMERA：照相机视频输入。
- DEFAULT：平台默认。

3. MediaRecorder 对象的编码方式

录音机接口支持的音频编码方式如下。

- AMR_NB：AMR 窄带。
- DEFAULT：默认编码。

（2）录像机接口支持的编码方式如下。

- H263：H.263 编码。
- H264：H.264 编码。
- MPEG_4_SP：MPEG4 编码。

4. MediaRecorder 对象的输出格式
- MPEG-4：MPEG4 格式。
- RAW_AMR：原始 AMR 格式文件。
- THREE_GPP：3GP 格式。

4.5.2 录音示例

应用 MediaRecorder 进行录音频，其主要步骤如下。

（1）创建录音对象。

```
MediaRecorder mRecorder = new MediaRecorder();
```

（2）录音对象的设置。
- 设置音频源：

```
mRecorder.setAudioSource(MediaRecorder.AudioSource.MIC);
```

- 设置输出格式：

```
mRecorder.setOutputFormat(MediaRecorder.OutputFormat.THREE_GPP);
```

- 设置编码格式：

```
mRecorder.setAudioEncoder(MediaRecorder.AudioEncoder.AMR_NB);
```

- 设置输出文件路径：

```
mRecorder.setOutputFile(path);
```

（3）准备录制。

```
mRecorder.prepare();
```

（4）开始录制。

```
mRecorder.start();
```

（5）结束录制。
- 停止录制：

```
mRecorder.stop();
```

- 重置：

```
mRecorder.reset();
```

- 释放录音占用的有关资源：

```
mRecorder.release();
```

【例 4-10】 设计一个简易录音机。

(1) 设计图形界面布局文件。在图形界面布局文件中设置两个按钮，一个按钮用于录音，另一个按钮用于停止录音。

(2) 设计控制文件 MainActivity.java。

```
1   package com.example.ex4_10;
2   import android.media.MediaRecorder;
3   import android.os.Bundle;
4   import android.app.Activity;
5   import android.view.View;
6   import android.view.View.OnClickListener;
7   import android.widget.Button;
8
9   public class MainActivity extends Activity
10  {
11    MediaRecorder mRecorder;
12    Button startBtn, stopBtn;
13      String path;
14      @Override
15      public void onCreate(Bundle savedInstanceState)
16      {
17        super.onCreate(savedInstanceState);
18        setContentView(R.layout.activity_main);
19        path = "/sdcard/zsm/audio.amr";
20        startBtn = (Button)findViewById(R.id.button1);
21        stopBtn = (Button)findViewById(R.id.button2);
22        startBtn.setOnClickListener(new mClick());
23        stopBtn.setOnClickListener(new mClick());
24      }
25
26    class mClick implements OnClickListener
27    {
28        @Override
29        public void onClick(View v)
30        {
31          if(v == startBtn)
32          {
33              startRecordAudio(path);
34          }
35          else if(v == stopBtn)
36          {
37              stopRecord();
38          }
39        }
40    }
41
42    void startRecordAudio(String path)
43    {
44        mRecorder=new MediaRecorder();
45        mRecorder.setAudioSource(MediaRecorder.AudioSource.MIC);   ← 设置音频源
```

```
46      mRecorder.setOutputFormat(
47              MediaRecorder.OutputFormat.THREE_GPP);    // 设置输出格式
48      mRecorder.setAudioEncoder(
49        MediaRecorder.AudioEncoder.AMR_NB);    // 设置编码格式
50      mRecorder.setOutputFile(path);    // 设置输出文件路径
51      try {
52          mRecorder.prepare();    // 录制准备
53        }catch (Exception e) {
54            System.out.println("Recorder err ... ");
55        }
56          mRecorder.start();    // 开始录音
57   }
58
59   void stopRecord()    // 停止录音
60   {
61      mRecorder.stop();    //停止录制
62      mRecorder.reset();    //重置
63      mRecorder.release();//释放播放器有关资源
64   }
65 }
```

（3）修改配置文件。在配置文件 AndroidManifest.xml 中要增加音频捕获权限的语句。

- 音频捕获权限：

`<uses-permission android:name="android.permission.RECORD_AUDIO"/>`

- SD 卡的写操作权限：

`<uses-permission android:name="android.permission.WRITE_EXTERNAL_STORAGE"/>`

程序的运行结果如图 4.14 所示。

图 4.14　录音程序运行界面

4.5.3 拍照

使用 android.hardware 包中的 Camera 类可以获取当前设备中的照相机服务接口，从而实现照相机的拍照功能。

1. 照片服务类 Camera

Camera 类的常用方法如表 4-9 所示。

表 4-9　Camera 类的常用方法

方法	说明
open()	创建一个照相机对象
getParameters()	创建设置照相机参数的 Camera.Parameters 对象
setParameters(Camera.Parameters params)	设置照相机参数
setPreviewDisplay(SurfaceHolder holder)	设置取景预览
startPreview()	启动照片取景预览
stopPreview()	停止照片取景预览
release()	断开与照相机设备的连接，并释放资源
takePicture(Camera.ShutterCallback shutter, Camera.PictureCallback raw, Camera.PictureCallback jpeg)	进行拍照

takePicture 方法有以下 3 个参数：

- 第 1 个参数 shutter 是关闭快门事件的回调接口。
- 第 2 个参数 raw 是获取照片事件的回调接口。
- 第 3 个参数 jpeg 也是获取照片事件的回调接口。

第 2 个参数与第 3 个参数的区别在于回调函数中传回的数据内容。第 2 个参数指定的回调函数中传回的数据内容是照片的原数据，而第 3 个参数指定的回调函数中传回的数据内容是已经按照 jpeg 格式进行编码的数据。

2. 实现拍照服务的主要步骤

在 Android 系统中实现拍照服务的主要步骤如下。

（1）创建照相机对象。通过 Camera 类的 open() 方法创建一个照相机对象：

```
Camera camera=Camera.open();
```

（2）设置参数。创建设置照相机参数的 Parameters 对象，并设置相关参数：

```
parameters = mCamera.getParameters();
```

（3）对照片预览。通过照相机对象的 startPreview() 方法和 stopPreview() 方法启动或停止对照片的预览。

（4）拍照。使用照相机接口的 takePicture() 方法可以异步地进行拍照。

通过照片事件的回调接口 PictureCallback 可以获取照相机所得到的图片数据，从而进行下一步的行动，例如保存到本地存储、进行数据压缩、通过可视组件显示。

（5）停止拍照。通过照相机对象的 release() 方法可以断开与照相机设备的连接，并释放与该照相机接口有关的资源。

```
camera.release();
camera=null;
```

【例 4-11】 设计一个简易照相机。

设计照相机,为了取景,需要应用 SurfaceView 组件来显示镜头所能拍照的景物,再使用回调接口 SurfaceHolder.Callback 监控取景视图,Callback 接口有 3 个方法需要实现。

- surfaceCreated(SurfaceHolder holder)方法:用于初始化。
- surfaceChanged(SurfaceHolder holder, int format, int width, int height)方法:当景物发生变化时触发。
- surfaceDestroyed(SurfaceHolder holder)方法:释放对象时触发。

(1)设计用户界面程序 main.xml。在界面设计中设置两个按钮,分别为"拍照"和"退出";再设置一个 SurfaceView 组件用于取景预览;再设置一个 ImageView 组件用于显示照片。其代码如下:

```
1  <LinearLayout xmlns:android="http://schemas.android.com/apk/res/android"
2      xmlns:tools="http://schemas.android.com/tools"
3      android:id="@+id/LinearLayout1"
4      android:layout_width="fill_parent"
5      android:layout_height="fill_parent"
6      android:orientation="vertical" >
7  <TextView
8      android:layout_width="wrap_content"
9      android:layout_height="wrap_content"
10     android:layout_gravity="center_horizontal"
11     android:padding="@dimen/padding_medium"
12     android:text="拍照测试"
13     android:textSize="20sp"
14     tools:context=".MainActivity" />
15 <LinearLayout
16     android:layout_width="fill_parent"
17     android:layout_height="wrap_content"
18     android:layout_gravity="center_horizontal"
19     android:gravity="center_horizontal" >
20     <Button
21         android:id="@+id/button1"
22         android:layout_width="110dp"
23         android:layout_height="wrap_content"
24         android:text="拍照" />
25     <Button
26         android:id="@+id/button2"
27         android:layout_width="110dp"
28         android:layout_height="wrap_content"
29         android:text="退出" />
30 </LinearLayout>
```

```xml
31      <ImageView          ← 用于显示拍照的相片
32          android:id="@+id/imageView1"
33          android:layout_width="wrap_content"
34          android:layout_height="wrap_content" />
35      <SurfaceView
36          android:id="@+id/surfaceView1"     ← 用于取景预览
37          android:layout_width="320dp"
38          android:layout_height="240dp" />
39  </LinearLayout>
```

（2）设计控制程序 MainActivity.java。

```java
1   package com.example.ex4_11;
2   import java.io.BufferedOutputStream;
3   import java.io.FileOutputStream;
4   import java.io.IOException;
5   import android.graphics.Bitmap;
6   import android.graphics.BitmapFactory;
7   import android.graphics.PixelFormat;
8   import android.hardware.Camera;
9   import android.hardware.Camera.PictureCallback;
10  import android.os.Bundle;
11  import android.util.Log;
12  import android.view.SurfaceHolder;
13  import android.view.SurfaceView;
14  import android.view.View;
15  import android.view.View.OnClickListener;
16  import android.widget.Button;
17  import android.widget.ImageView;
18  import android.app.Activity;
19
20  public class MainActivity extends Activity
21          implements SurfaceHolder.Callback     ← 实现 Callback 接口处理取景预览
22  {
23    Camera mCamera=null;
24    SurfaceView surfaceView;
25    SurfaceHolder holder;
26    ImageView mImageView;
27    Button cameraBtn, exitBtn;
28    String path = "/sdcard/test/camera.jpg";   ← 定义存放相片的路径及文件名
29    @Override
30    public void onCreate(Bundle savedInstanceState)
31    {
32        super.onCreate(savedInstanceState);
33        setContentView(R.layout.activity_main);
34        mImageView = (ImageView)findViewById(R.id.imageView1);
```

```java
35        cameraBtn = (Button)findViewById(R.id.button1);
36        exitBtn = (Button)findViewById(R.id.button2);
37        cameraBtn.setOnClickListener(new mClick());
38        exitBtn.setOnClickListener(new mClick());
39        surfaceView =(SurfaceView)findViewById(R.id.surfaceView1);
40        //创建SurfaceHolder对象
41        holder = surfaceView.getHolder();
42        //注册回调监听器
43        holder.addCallback(this);
44        //设置SurfaceHolder的类型
45        holder.setType(SurfaceHolder.SURFACE_TYPE_PUSH_BUFFERS);
46    }
47    class mClick implements OnClickListener
48    {
49        @Override
50        public void onClick(View v)
51        {
52            if(v == cameraBtn)
53                /* 拍照并显示相片 */
54                mCamera.takePicture(null, null, new jpegCallback());    ← 拍照操作
55            else if(v == exitBtn)
56                exit();    ← 退出
57        }
58    }
59    void exit()
60    {
61        mCamera.release();
62        mCamera = null;
63    }
64    @Override
65    public void surfaceChanged(SurfaceHolder holder,    ← 当景物发生变化时触发
66                int format, int width, int height)
67    {
68    /* 调用设置照相机取景参数的方法 */
69    initCamera();
70    }
71    @Override
72    public void surfaceCreated(SurfaceHolder holder)
73    {
74    /* 打开相机 */
75        mCamera = Camera.open();
76        try {
77            /* 设置预览 */
78            mCamera.setPreviewDisplay(holder);
79        } catch (IOException e) {
```

```
80          System.out.println("预览错误");
81      }
82  }
83  @Override
84  public void surfaceDestroyed(SurfaceHolder holder)
85  {           }
86  /* 设置照相机取景参数 */
87  private void initCamera()
88  {
89      /* 创建Camera.Parameters对象 */
90      Camera.Parameters parameters = mCamera.getParameters();
91      /* 设置相片为jpeg格式 */
92      parameters.setPictureFormat(PixelFormat.JPEG);
93      /* 指定preview的屏幕大小 */
94      parameters.setPreviewSize(320, 240);
95      /* 设置图片的分辨率大小 */
96      parameters.setPictureSize(320, 240);
97      /* 将Camera.Parameters的设置作用于Camera对象 */
98      mCamera.setParameters(parameters);
99      /* 打开预览 */
100     mCamera.startPreview();        ← 取景预览,此时还没有拍照保存为相片
101 }
102 /* 通过PictureCallback接口进一步处理照相机所得到的图像数据 */
103 class jpegCallback implements PictureCallback
104 {
105     /** 下面onPictureTaken()方法将图像转换成jpg格式后保存并预览,
106      *  其中第1个参数data为存放相片数据的字节数组,
107      *  第2个参数camera为相片对象 */
108     @Override
109     public void onPictureTaken(byte[] data, Camera camera)
110     {
111         Bitmap bitmap =
112             BitmapFactory.decodeByteArray(data, 0, data.length);   ← 建立图像对象
113         try{
114         BufferedOutputStream outStream = new
115             BufferedOutputStream(new FileOutputStream(path));   ← 建立输出流对象
116         /* 采用压缩转档方法 */
117         bitmap.compress(Bitmap.CompressFormat.JPEG, 80, outStream);
118         outStream.flush();    ← 调用 flush()方法,更新 BufferStream
119         outStream.close();
120         /* 显示拍照的图像 */
121         mImageView.setImageBitmap(bitmap);
122         }
123         catch (Exception e)
124         {
```

```
125            Log.e("err", e.getMessage());
126        }
127    }
128 }
129 }
```

(3)修改配置文件 AndroidManifest.xml。在配置文件 AndroidManifest.xml 中增加允许操作 SD 卡和使用镜头设备的语句:

```
<uses-permission android:name="android.permission.CAMERA"/>
<uses-permission android:name="android.permission.WRITE_EXTERNAL_STORAGE"/>
    <uses-feature android:name="android.hardwre.camera"/>
    <uses-feature android:name="android.hardwre.camera.autofocus"/>
```

程序的运行结果如图 4.15 所示。

图 4.15 简易照相机

4.6 动画技术

4.6.1 动画组件类

1. 动画组件 Animations 概述

Animations 是一个实现 Android UI 界面动画效果的 API,Animations 提供了一系列的动画效果,可以进行旋转、缩放、淡入淡出等,这些效果可以应用在绝大多数的控件中。

2. 动画组件 Animations 的分类

Animations 从总体上可以分为以下两大类:

1）补间动画

补间动画（Tween Animation）就是只需指定开始、结束的"关键帧"，而变化中的其他帧由系统来计算，不必一帧一帧地去定义。

2）逐帧动画

逐帧动画（Frame Animation）可以创建一个 Drawable 序列，这些 Drawable 可以按照指定的时间间隔一个一个显示。

3）属性动画

属性动画（Property Animation）是在 Android 3.0 版本以后才引进的，它可以直接更改对象的属性。在上面提到的 Tween Animation 中只是更改 View 的绘画效果，而 View 的真实属性是不改变的。假设用 Tween Animation 动画将一个 Button 从左边移到右边，无论怎么点击移动后的 Button 都没有反应，而当点击移动前 Button 的位置时才有反应，因为 Button 的位置属性没有改变。Property Animation 属性动画则可以直接改变 View 对象的属性值，这样可以让编程人员少做一些处理工作，提高效率与代码的可读性。

4.6.2 补间动画 Tween Animation

1. 补间动画效果的种类及对应的子类

补间动画（Tween Animation）共有 4 种动画效果及对应的子类。

（1）Alpha：淡入淡出效果，其对应子类为 AlphaAnimation。

（2）Scale：缩放效果，其对应子类为 ScaleAnimation。

（3）Rotate：旋转效果，其对应子类为 RotateAnimation。

（4）Translate：移动效果，其对应子类为 TranslateAnimation。

动画效果对应子类的构造方法如表 4-10 所示。

表 4-10 动画效果对应子类的构造方法

对应子类的构造方法	参数说明
AlphaAnimation (　　float fromAlpha, 　　float toAlpha)	参数 1 fromAlpha：起始透明度 参数 2 toAlpha：终止透明度 （取 0.0～1.0 的数值，1.0 为完全不透明，0.0 为完全透明）
ScaleAnimation(　　float fromX, 　　float toX, 　　float fromY, 　　float toY, 　　int pivotXType, 　　float pivotXValue, 　　int pivotYType, 　　float pivotYValue)	参数 1 fromX：X 轴的初始值 参数 2 toX：X 轴收缩后的值 参数 3 fromY：Y 轴的初始值 参数 4 toY：Y 轴收缩后的值 参数 5 pivotXType：确定 X 轴坐标的类型 参数 6 pivotXValue：X 轴的值，0.5f 表明是以自身这个控件的一半长度为 X 轴 参数 7 pivotYType：确定 Y 轴坐标的类型 参数 8 pivotYValue：Y 轴的值，0.5f 表明是以自身这个控件的一半长度为 X 轴

续表

对应子类的构造方法	参数说明

RotateAnimation(
    float fromDegrees,
    float toDegrees,
    int pivotXType,
    float pivotXValue,
    int pivotYType,
    float pivotYValue
)
	参数1 fromDegrees：从哪个旋转角度开始
	参数2 toDegrees：转到什么角度
	后面4个参数用于设置围绕着旋转的圆的圆心位置
	参数3 pivotXType：确定X轴坐标的类型，有ABSOLUT 绝对坐标、RELATIVE_TO_SELF 相对于自身坐标、RELATIVE_TO_PARENT 相对于父控件的坐标
	参数4 pivotXValue：X轴的值，0.5f 表明是以自身这个控件的一半长度为X轴
	参数5 pivotYType：确定Y轴坐标的类型
	参数6 pivotYValue：Y轴的值，0.5f 表明是以自身这个控件的一半长度为 x 轴

TranslateAnimation(
    float fromXDelta,
    float toXDelta,
    float fromYDelta,
    float toYDelta
)
	参数1 fromXDelta：X轴的开始位置
	参数2 toXDelta：X轴的结束位置
	参数3 fromYDelta：Y轴的开始位置
	参数4 toYDelta：X轴的结束位置

2. AnimationSet 类

AnimationSet 类是 Animation 的子类，用于设置 Animation 的属性。

【例 4-12】 编写一个可以旋转、缩放、淡入淡出、移动的补间动画程序。

（1）新建工程 ex4_12，并将事先准备的图片 an.jpg 复制到项目的 res\drawable 目录下，如图 4.16 所示。

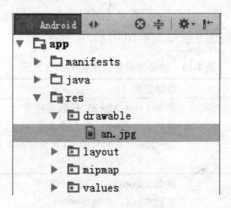

图 4.16 将图片 an.jpg 复制到项目的 res\drawable 目录下

（2）编写界面布局 XML 程序。

在界面布局 XML 程序中按垂直线性布局，并放置 4 个按钮组件 Button 和一个图像显示组件 ImageView。

程序如下：

```
1  <?xml version="1.0" encoding="utf-8"?>
```

```xml
2  <LinearLayout xmlns:android="http://schemas.android.com/apk/res/android"
3      xmlns:tools="http://schemas.android.com/tools"
4      android:layout_width="fill_parent"
5      android:layout_height="fill_parent"
6      android:orientation="vertical"
7      tools:context="com.example.ex4_12.MainActivity">
8      <TextView
9          android:layout_width="wrap_content"
10         android:layout_height="wrap_content"
11         android:text="动画演示"
12         android:textSize="26sp"
13         android:layout_gravity="center_horizontal" />
14     <LinearLayout
15         android:layout_width="fill_parent"
16         android:layout_height="fill_parent"
17         android:orientation="vertical" >
18         <LinearLayout
19             android:layout_width="wrap_content"
20             android:layout_height="wrap_content"
21             android:orientation="horizontal" >
22             <Button
23                 android:id="@+id/rotateButton"
24                 android:layout_width="wrap_content"
25                 android:layout_height="wrap_content"
26                 android:text="旋转" />
27             <Button
28                 android:id="@+id/scaleButton"
29                 android:layout_width="wrap_content"
30                 android:layout_height="wrap_content"
31                 android:text="缩放" />
32             <Button
33                 android:id="@+id/alphaButton"
34                 android:layout_width="wrap_content"
35                 android:layout_height="wrap_content"
36                 android:text="淡入淡出" />
37             <Button
38                 android:id="@+id/translateButton"
39                 android:layout_width="wrap_content"
40                 android:layout_height="wrap_content"
41                 android:text="移动" />
42         </LinearLayout>
43         <ImageView
44             android:id="@+id/image"
45             android:layout_width="wrap_content"
```

```
46            android:layout_height="wrap_content"
47            android:layout_centerInParent="true"
48            android:src="@drawable/an"
49            android:layout_gravity="center" />
50      </LinearLayout>
51 </LinearLayout>
```

(3) 编写控制层 Java 程序。

```
1  package com.example.ex4_12;
2  import android.support.v7.app.AppCompatActivity;
3  import android.os.Bundle;
4  import android.view.View;
5  import android.view.animation.AlphaAnimation;
6  import android.view.animation.Animation;
7  import android.view.animation.AnimationSet;
8  import android.view.animation.RotateAnimation;
9  import android.view.animation.ScaleAnimation;
10 import android.view.animation.TranslateAnimation;
11 import android.widget.Button;
12 import android.widget.ImageView;
13 import android.view.View.OnClickListener;
14 public class MainActivity extends AppCompatActivity {
15     private Button rotateButton = null;
16     private Button scaleButton = null;
17     private Button alphaButton = null;
18     private Button translateButton = null;
19     private ImageView image = null;
20     @Override
21     protected void onCreate(Bundle savedInstanceState) {
22         super.onCreate(savedInstanceState);
23         setContentView(R.layout.activity_main);
24         rotateButton = (Button)findViewById(R.id.rotateButton);
25         scaleButton = (Button)findViewById(R.id.scaleButton);
26         alphaButton = (Button)findViewById(R.id.alphaButton);
27         translateButton = (Button)findViewById(R.id.translateButton);
28         image = (ImageView)findViewById(R.id.image);
29         rotateButton.setOnClickListener(new RotateButtonListener());
30         scaleButton.setOnClickListener(new ScaleButtonListener());
31         alphaButton.setOnClickListener(new AlphaButtonListener());
32         translateButton.setOnClickListener(new TranslateButtonListener());
33     }
34     class RotateButtonListener implements OnClickListener{
35         public void onClick(View v) {
36             AnimationSet animationSet = new AnimationSet(true);
```

```
37          RotateAnimation rotateAnimation = new RotateAnimation(0, 360,
38                  Animation.RELATIVE_TO_SELF, 0.5f,
39                  Animation.RELATIVE_TO_SELF, 0.5f);
40          rotateAnimation.setDuration(1000);
41          animationSet.addAnimation(rotateAnimation);
42          image.startAnimation(animationSet);
43      }
44  }
45  class ScaleButtonListener implements OnClickListener{
46      public void onClick(View v) {
47          AnimationSet animationSet = new AnimationSet(true);
48          ScaleAnimation scaleAnimation = new ScaleAnimation(
49                  0, 0.1f, 0, 0.1f, Animation.RELATIVE_TO_SELF,
50                  0.5f, Animation.RELATIVE_TO_SELF, 0.5f);
51          scaleAnimation.setDuration(1000);
52          animationSet.addAnimation(scaleAnimation);
53          image.startAnimation(animationSet);
54      }
55  }
56  class AlphaButtonListener implements OnClickListener{
57      public void onClick(View v) {
58          //创建一个AnimationSet对象，参数为Boolean型
59          AnimationSet animationSet = new AnimationSet(true);
60          //创建一个AlphaAnimation对象，参数从完全不透明度到完全透明
61          AlphaAnimation alphaAnimation = new AlphaAnimation(1, 0);
62          //设置动画执行的时间
63          alphaAnimation.setDuration(500);
64          //将alphaAnimation对象添加到AnimationSet中
65          animationSet.addAnimation(alphaAnimation);
66          //使用ImageView的startAnimation方法执行动画
67          image.startAnimation(animationSet);
68      }
69  }
70  class TranslateButtonListener implements OnClickListener{
71      public void onClick(View v) {
72          AnimationSet animationSet = new AnimationSet(true);
73          TranslateAnimation translateAnimation =
74                  new TranslateAnimation(
75                          Animation.RELATIVE_TO_SELF, 0f,
76                          Animation.RELATIVE_TO_SELF, 0.5f,
77                          Animation.RELATIVE_TO_SELF, 0f,
78                          Animation.RELATIVE_TO_SELF, 0.5f);
79          translateAnimation.setDuration(1000);
80          animationSet.addAnimation(translateAnimation);
81          image.startAnimation(animationSet);
```

```
82        }
83    }
84 }
```

程序的运行结果如图 4.17 所示。

图 4.17 补间动画示例

4.6.3 属性动画 Property Animation

1. 属性动画的核心类

属性动画就是通过控制对象中的属性值产生的动画。属性动画主要的核心类有 ValueAnimator 和 ObjectAnimator。

1）ValueAnimator 类

ValueAnimator 是整个属性动画机制当中最核心的一个类。属性动画的运行机制是通过不断地对值进行操作来实现的,而初始值和结束值之间的动画过渡是由 ValueAnimator 这个类来负责计算的。它的内部使用一种时间循环的机制来计算值与值之间的动画过渡,用户只需要将初始值和结束值提供给 ValueAnimator,并且告诉它动画所需运行的时长,那么 ValueAnimator 就会自动帮助用户完成从初始值平滑地过渡到结束值。除此之外,ValueAnimator 还负责管理动画的播放次数、播放模式以及对动画设置监听器等。

2）ObjectAnimator 类

ObjectAnimator 是 ValueAnimator 的子类,它本身就已经包含了时间引擎和值计算,所以它拥有为对象的某个属性设置动画的功能,这使得为任何对象设置动画更加容易。

ObjectAnimator 类是设计动画时最常使用的类,ValueAnimator 只不过是对值进行了一个平滑的动画过渡,所以实际应用到这种功能的场景并不多。而 ObjectAnimator 就不同了,它是可以直接对任意对象的任意属性进行动画操作的，比如 View 的 alpha 属性。

构造 ObjectAnimator 对象的方法为 ofFloat()，其方法原型如下：

```
public static ObjectAnimator ofFloat(
    Object target,
    String propertyName,
    float... values
    );
```

- 第 1 个参数用于指定动画对象要操作的控件。
- 第 2 个参数用于指定动画对象所要操作控件的属性。
- 第 3 个参数是可变长参数，设置动画的起点和终点位置。

2. 应用 ObjectAnimator 类实现动画示例

下面通过一个示例来说明如何应用 ObjectAnimator 类实现动画。

【例 4-13】 编写一个可以旋转、缩放、淡入淡出的属性动画程序。

（1）新建工程 ex4_13，并将事先准备的图片 an.jpg 复制到项目的 res\drawable 目录下。

（2）编写界面布局 XML 程序。在界面布局 XML 程序中按垂直线性布局，并放置按钮组件 Button 和图像显示组件 ImageView。

程序如下：

```
1  <?xml version="1.0" encoding="utf-8"?>
2  <LinearLayout xmlns:android="http://schemas.android.com/apk/res/android"
3    xmlns:tools="http://schemas.android.com/tools"
4    android:layout_width="fill_parent"
5    android:layout_height="fill_parent"
6    android:orientation="vertical"
7    tools:context="com.example.ex4_13.MainActivity">
8    <TextView
9        android:layout_width="wrap_content"
10       android:layout_height="wrap_content"
11       android:text="属性动画演示"
12       android:textSize="26sp"
13       android:layout_gravity="center_horizontal" />
14   <LinearLayout
15       android:layout_width="fill_parent"
16       android:layout_height="fill_parent"
17       android:orientation="vertical" >
18       <LinearLayout
19           android:layout_width="wrap_content"
20           android:layout_height="wrap_content"
21           android:orientation="horizontal" >
22           <Button
23               android:id="@+id/rotateButton"
24               android:layout_width="wrap_content"
25               android:layout_height="wrap_content"
26               android:text="旋转" />
```

```
27        <Button
28            android:id="@+id/scaleButton"
29            android:layout_width="wrap_content"
30            android:layout_height="wrap_content"
31            android:text="缩放" />
32        <Button
33            android:id="@+id/alphaButton"
34            android:layout_width="wrap_content"
35            android:layout_height="wrap_content"
36            android:text="淡入淡出" />
37        <Button
38            android:id="@+id/translateButton"
39            android:layout_width="wrap_content"
40            android:layout_height="wrap_content"
41            android:text="移动" />
42    </LinearLayout>
43    <ImageView
44        android:id="@+id/image"
45        android:layout_width="wrap_content"
46        android:layout_height="wrap_content"
47        android:layout_centerInParent="true"
48        android:src="@drawable/bn"
49        android:layout_gravity="center" />
50   </LinearLayout>
51 </LinearLayout>
```

(3) 编写控制层 Java 程序。

```
1  package com.example.ex4_13;
2  import （略）
3
4  public class MainActivity extends AppCompatActivity {
5      Button rotateButton,alphaButton,scaleButton;
6      ImageView img;
7      @Override
8      protected void onCreate(Bundle savedInstanceState) {
9          super.onCreate(savedInstanceState);
10         setContentView(R.layout.activity_main);
11         img = (ImageView)findViewById(R.id.imageView);
12         rotateButton = (Button)findViewById(R.id.button1);
13         alphaButton = (Button)findViewById(R.id.button2);
14         scaleButton = (Button)findViewById(R.id.button3);
15         rotateButton.setOnClickListener(new mClick());
16         alphaButton.setOnClickListener(new mClick());
17         scaleButton.setOnClickListener(new mClick());
18     }
19     public class mClick implements View.OnClickListener
```

```
20    {
21        @Override
22        public void onClick(View v) {
23            if(v == rotateButton) {
24                ObjectAnimator animator = ObjectAnimator.ofFloat(img,
                                                "rotation", 0.0F, 360.0F);
25                animator.setDuration(1000);
26                animator.start();
27            }
28            else if(v == alphaButton){
29                ObjectAnimator animator = ObjectAnimator.ofFloat(img,
                                                "alpha",1.0F, 0.0F, 1.0F);
30                animator.setDuration(3000);
31                animator.start();
32            }
33            else if(v == scaleButton){
34                ObjectAnimator animator = ObjectAnimator.ofFloat(img,
                                                "ScaleY", 1.0F, 0.5F, 1.0F);
35                animator.setDuration(5000);
36                animator.start();
37            }
38        }
39    }
40 }
```

程序的运行结果如图 4.18 所示。

图 4.18　属性动画示例

习 题 4

1. 设计一个可以移动的小球，当小球被拖到一个小矩形块中时退出程序。
2. 设计一个手绘图形的画板。
3. 建立一个手写字体识别的字体库。
4. 设计一个具有选歌功能的音频播放器。
5. 为例 4-8 的视频播放器添加停止播放的功能。
6. 编写一个具有飞入文字功能的程序。

第 5 章　后台服务与系统服务

5.1　后台服务 Service

Android 系统的 Service 是一种类似于 Activity 的组件，但 Service 没有用户操作界面，也不能自己启动，其主要作用是提供后台服务调用。Service 不像 Activity 那样当用户关闭应用界面时就停止运行，Service 会一直在后台运行，除非明确命令其停止。

通常使用 Service 为应用程序提供一些只需在后台运行的服务或不需要界面的功能，例如从 Internet 下载文件、控制 Video 播放器等。

在 Service 的生命周期中只有 3 个阶段，即 onCreate、onStartCommand、onDestroy。其生命周期的方法见表 5-1。

表 5-1　Service 服务的常用方法

方法	说明
onCreate()	创建后台服务
onStartCommand (Intent intent, int flags, int startId)	启动一个后台服务
onDestroy()	销毁后台服务，并删除所有调用
sendBroadcast(Intent intent)	继承父类 Context 的 sendBroadcast()方法，实现发送广播机制的消息
onBind(Intent intent)	与服务通信的信道进行绑定，服务程序必须实现该方法
onUnbind(Intent intent)	撤销与服务信道的绑定

通常 Service 要在一个 Activity 中启动，调用 Activity 的 startService(Intent)方法启动 Service。若要停止正在运行的 Service，则调用 Activity 的 stopService(Intent)方法关闭 Service。startService()和 stopService()方法均继承于 Activity 及 Service 共同的父类 android.content.Context。

一个服务只能创建一次，销毁一次，但可以开始多次，即 onCreate()和 onDestroy()方法只会被调用一次，而 onStartCommand()方法可以被调用多次。后台服务的具体操作一般应该放在 onStartCommand()方法里面。如果 Service 已经启动，当再次启动 Service 时则不调用 onCreate()而直接调用 onStartCommand()。

设计一个后台服务的应用程序大致有以下几个步骤。

（1）创建 Service 的子类：
- 编写 onCreate()方法，创建后台服务；
- 编写 onStartCommand()方法，启动后台服务；

- 编写 onDestroy()方法，终止后台服务，并删除所有调用。

（2）创建启动和控制 Service 的 Activity：
- 创建 Intent 对象，建立 Activity 与 Service 的关联；
- 调用 Activity 的 startService(Intent)方法启动 Service 后台服务；
- 调用 Activity 的 stopService(Intent)方法关闭 Service 后台服务。

（3）修改配置文件 AndroidManifest.xml。在配置文件 AndroidManifest.xml 的<application>标签中添加以下代码：

```
<service android:enabled="true" android:name=".AudioSrv" />
```

【例 5-1】 一个简单的后台音乐服务程序示例。

本例通过一个按钮启动后台服务，在服务程序中播放音乐文件，演示服务程序的创建、启动，再通过另一按钮演示服务程序的销毁过程。新建项目 ex5_1 后将音频文件 mtest1.mp3 复制到应用程序的资源目录 res\raw 下。

（1）设计界面布局 activity_main.xml。

```
1   <?xml version="1.0" encoding="utf-8"?>
2   <LinearLayout xmlns:android="http://schemas.android.com/apk/res/android"
3       android:layout_width="fill_parent"
4       android:layout_height="fill_parent"
5       android:orientation="vertical" >
6       <TextView
7           android:id="@+id/text1"
8           android:layout_width="fill_parent"
9           android:layout_height="wrap_content"
10          android:text="@string/hello"
11          android:textSize="24sp"/>
12      <Button
13          android:id="@+id/butn1"
14          android:layout_width="wrap_content"
15          android:layout_height="wrap_content"
16          android:text="启动后台音乐服务程序"
17          android:textSize="24sp" />
18      <Button
19          android:id="@+id/butn2"
20          android:layout_width="wrap_content"
21          android:layout_height="wrap_content"
22          android:text="关闭后台音乐服务程序"
23          android:textSize="24sp" />
24  </LinearLayout>
```

（2）新建后台服务程序 AudioSrv.java。

```
1   package com.ex5_1;
```

```java
2  import android.app.Service;
3  import android.content.Intent;
4  import android.media.MediaPlayer;
5  import android.os.IBinder;
6  import android.widget.Toast;
7
8  public class AudioSrv extends Service
9  {
10 MediaPlayer play;
11 @Override
12 public IBinder onBind(Intent intent)
13 {
14     return null;
15 }
16 public void onCreate()
17 {
18   super.onCreate();
19   play=MediaPlayer.create(this, R.raw.mtest1);   ← 创建调用资源音乐文件对象
20   Toast.makeText(this, "创建后台服务...", Toast.LENGTH_LONG).show();
21 }
22 public int onStartCommand(Intent intent, int flags, int startId)
23 {
24     super.onStartCommand(intent, flags, startId);
25     play.start();   ← 开始播放音乐
26     Toast.makeText(this, "启动后台服务程序，播放音乐...",
27                     Toast.LENGTH_LONG).show();
28     return START_STICKY;
29 }
30 public void onDestroy()
31 {
32   play.release();
33   super.onDestroy();
34   Toast.makeText(this, "销毁后台服务!", Toast.LENGTH_LONG).show();
35 }
36 }
```

（3）启动后台服务的主控程序 MainActivity.java。

```java
1  package com.ex5_1;
2  import android.app.Activity;
3  import android.content.Context;
4  import android.content.Intent;
5  import android.os.Bundle;
6  import android.view.View;
7  import android.view.View.OnClickListener;
8  import android.widget.Button;
```

```
9    import android.widget.TextView;
10
11   public class MainActivity extends Activity
12   {
13   Button startbtn, stopbtn;
14   Context context;
15   Intent intent;
16   static TextView txt;
17    @Override
18    public void onCreate(Bundle savedInstanceState)
19    {
20      super.onCreate(savedInstanceState);
21      setContentView(R.layout.main);
22      startbtn=(Button)findViewById(R.id.butn1);
23      stopbtn=(Button)findViewById(R.id.butn2);
24      startbtn.setOnClickListener(new mClick());
25      stopbtn.setOnClickListener(new mClick());
26      txt=(TextView)findViewById(R.id.text1);
27      intent=new Intent(MainActivity.this, AudioSrv.class);   ← 创建 Intent 对象
28    }
29    class mClick implements OnClickListener    //定义一个类实现监听接口
30    {
31      public void onClick(View v)
32      {
33       if(v == startbtn)
34       {
35         MainActivity.this.startService(intent);   ← 启动 Intent 关联的 Service
36         txt.setText("start service .......");
37       }
38       else if(v == stopbtn)
39       {
40         MainActivity.this.stopService(intent);   ← 终止后台服务
41       }
42      }
43    }
44   }
```

（4）修改配置文件 AndroidManifest.xml。

在配置文件 AndroidManifest.xml 的<application>标签中添加以下代码：

```
<service android:enabled="true" android:name=".AudioSrv" />
```

修改后完整的 AndroidManifest.xml 文件如下：

```
1   <?xml version="1.0" encoding="utf-8"?>
2   <manifest xmlns:android="http://schemas.android.com/apk/res/android"
```

```
3       package="com.ex5_1"
4       android:versionCode="1"
5       android:versionName="1.0" >
6       <uses-sdk android:minSdkVersion="15" />
7       <application
8           android:icon="@drawable/ic_launcher"
9           android:label="@string/app_name" >
10          <activity
11              android:name=".MainActivity"
12              android:label="@string/app_name" >
13              <intent-filter>
14                  <action android:name="android.intent.action.MAIN" />
15                  <category android:name="android.intent.category.LAUNCHER" />
16              </intent-filter>
17          </activity>
18          <!-- 添加AudioSrv服务程序 -->
19          <service android:enabled="true" android:name=".AudioSrv" />
20      </application>
21  </manifest>
```

程序的运行结果如图 5.1 所示。

图 5.1　启动后台服务程序

5.2　信息广播机制 Broadcast

Broadcast 是 Android 系统应用程序之间传递信息的一种机制。当系统之间需要传递某些信息时不是通过诸如单击按钮之类的组件来触发事件, 而是由系统自身通过系统调用来引发事件。这种系统调用是由 BroadcastReceiver 类实现的,把这种系统调用称为广播。BroadcastReceiver 是"广播接收者"的意思,顾名思义,它用来接收来自系统和应用中的广播信息。

在 Android 系统中有很多广播信息,例如当开机时系统会产生一条广播信息,接收到这条广播信息就能实现开机启动服务的功能;当网络状态改变时系统会产生一条广播信息,接收到这条广播信息就能及时地做出提示和保存数据等操作;当电池电量改变时系统会产生一条广播信息,接收到这条广播信息就能在电量低时告知用户及时保存进度,等等。

实现广播和接收机制有以下 5 个步骤:

(1)创建 Intent 对象,设置 Intent 对象的 action 属性。这个 action 属性是接收广播数据的标识,只有注册了相同 action 属性的广播接收器才能收到发送的广播数据。

```
Intent intent = new Intent();
intent.setAction("abc");      ◀── 设置 Intent 对象的 action 属性值为"abc"
```

(2)编写需要广播的信息内容,将需要播发的信息封装到 Intent 中,通过 Activity 或 Service 继承其父类 Context 的 sendBroadcast()方法将 Intent 广播出去。

```
intent.putExtra("hello", "这是广播信息!");   ◀── 以键-值对形式封装广播信息内容
    sendBroadcast(intent);
```

(3)编写一个继承 BroadcastReceiver 的子类作为广播接收器,该对象是接收广播信息并对信息进行处理的组件。在子类中要重写接收广播信息的 onReceive()方法。

```
class TestReceiver extends BroadcastReceiver
    {
        @Override
        public void onReceive(Context context, Intent intent)
{
    /*   接收广播信息并对信息做出响应的代码   */
}
    }
```

(4)在配置文件 AndroidManifest.xml 中注册广播接收类。

```
    <service android:name=".TestReceiver">   ◀── 注册广播接收类
    <intent-filter>
      <action android:name="abc" />   ◀── action 属性值相同才能接收到广播数据
    </intent-filter>
  </service>
```

(5)销毁广播接收器。Android 系统在执行 onReceive()方法时会启动一个程序计时器,在一定时间内广播接收器的实例会被销毁。因此广播机制不适合传递大数据量的信息。

【例 5-2】 一个简单的信息广播程序示例。

(1)设计主控文件 MainActivity.java。

```
1  package com.ex5_2;
2  import android.app.Activity;
3  import android.content.Intent;
4  import android.os.Bundle;
5  import android.view.View;
6  import android.view.View.OnClickListener;
7  import android.widget.Button;
8  import android.widget.TextView;
9
10 public class MainActivity extends Activity
11 {
12   static TextView txt;
13   @Override
14   public void onCreate(Bundle savedInstanceState)
15   {
16     super.onCreate(savedInstanceState);
17     setContentView(R.layout.main);
18     txt = (TextView)findViewById(R.id.txt1);
19     Button btn=(Button)findViewById(R.id.button01);
20     btn.setOnClickListener(new mClick());
21   }
22
23   class mClick implements OnClickListener
24   {
25     @Override
26     public void onClick(View v)
27     {
28       Intent intent = new Intent();
29       intent.setAction("abc");         ← 设置action属性值
30       Bundle bundle = new Bundle();
31       bundle.putString("hello", "这是广播信息!"); ← 设置广播的消息内容
32       intent.putExtras(bundle);
33       sendBroadcast(intent);           ← 发送广播消息
34     }
35   }
36 }
```

(2) 设计广播接收器。

```
1  package com.ex5_2;
2  import android.content.BroadcastReceiver;
3  import android.content.Context;
4  import android.content.Intent;
5
6  public class TestReceiver extends BroadcastReceiver   ← 定义广播接收器
7  {
```

```
8       @Override
9       public void onReceive(Context context, Intent intent)
10      {
11        String str = intent.getExtras().getString("hello");   ← 取出接收的数据
12        MainActivity.txt.setText(str);    ← 显示接收的数据
13      }
14  }
```

（3）设计配置文件 AndroidManifest.xml。

```
1  <?xml version="1.0" encoding="utf-8"?>
2  <manifest xmlns:android="http://schemas.android.com/apk/res/android"
3    package="com.ex5_2"
4    android:versionCode="1"
5    android:versionName="1.0" >
6    <uses-sdk android:minSdkVersion="15" />
7    <application
8      android:icon="@drawable/ic_launcher"
9      android:label="@string/app_name" >
10     <activity
11       android:name=".MainActivity"
12       android:label="@string/app_name" >
13       <intent-filter>
14         <action android:name="android.intent.action.MAIN" />
15         <category android:name="android.intent.category.LAUNCHER" />
16       </intent-filter>
17     </activity>
18     <!-- 注册对应的广播接收类 -->
19     <receiver  android:name=".TestReceiver">
20       <intent-filter>
21         <!-- 注册广播的action，与setAction()设置的值相同 -->
22         <action android:name="abc" />
23       </intent-filter>
24     </receiver>
25   </application>
26 </manifest>
```

程序的运行结果如图 5.2 所示。

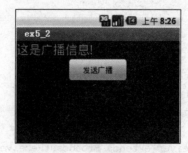

图 5.2 简单的广播示例

为了识别 Intent 对象的 action，有时在 IntentFilter 对象中设置 Intent 对象的 action，而注册广播接收器的工作由 registerReceiver()方法完成。

registerReceiver(mBroadcast, filter)方法有两个参数，其中参数 mBroadcast 是广播接收器对象 BroadcastReceiver，filter 是 IntentFilter 对象。

【例 5-3】 由一个后台服务广播音乐的播放或暂停信息，接收器接收到信息后执行改变用户界面按钮上文本的操作。

在本例中创建了 3 个类，即 MainActivity、AudioService 和 Broadcast，MainActivity 负责用户的交互界面，并启动后台服务；AudioService 是 Service 的子类，在后台提供播放音乐或暂停、停止音乐等工作，同时发送改变交互界面的广播信息；Broadcast 是 BroadcastReceiver 的子类，负责接收广播信息，更改交互界面。这 3 个类的工作流程如图 5.3 所示。

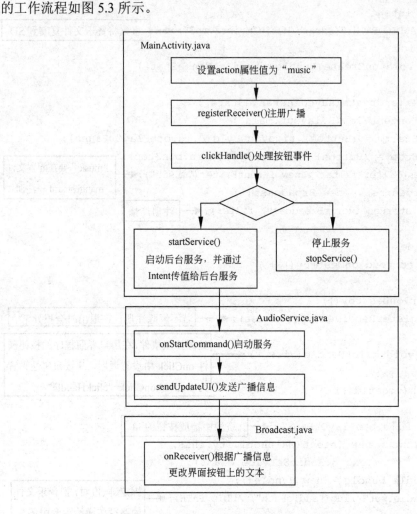

图 5.3 工作流程

（1）设计主控文件 MainActivity.java。

```
1   package com.ex5_3;
```

```java
2   import android.app.Activity;
3   import android.content.Intent;
4   import android.content.IntentFilter;
5   import android.os.Bundle;
6   import android.view.View;
7   import android.widget.Button;
8
9   public class MainActivity extends Activity
10  {
11    Broadcast mBroadcast = null;
12     static Button btnStart;
13    Button btnStop;
14    Intent intent;
15    String AUDIO_PATH="/sdcard/Music/mtest2.mp3";   ← 事先将音乐文件复制到 SD
16     @Override
17    public void onCreate(Bundle savedInstanceState)
18    {
19       super.onCreate(savedInstanceState);
20       setContentView(R.layout.main);
21       btnStart = (Button) findViewById(R.id.btnPlayOrPause);
22       btnStop = (Button) findViewById(R.id.btnStop);
23       IntentFilter filter = new IntentFilter("music");   ← "music" 要在配置文件
24        mBroadcast = new Broadcast();                        manifest.xml 里注册
25       registerReceiver(mBroadcast, filter);   ← 注册广播
26    }
27    @Override
28    protected void onDestroy()
29    {
30       super.onDestroy();
31       unregisterReceiver(mBroadcast);   ← 若不解除注册，在退出时会报异常
32    }
33    public void clickHandle(View v)   ← 处理按钮事件,在用户界面程序的按钮属
34    {                                    性 onClick 中设置调用，其按钮属性设置
35      switch(v.getId())                  为 android:onClick="clickHandle"
36      {
37      case R.id.btnPlayOrPause:   ← 资源中播放按钮的 id
38        intent = new Intent(MainActivity.this,
39             com.ex5_3.AudioService.class);
40        Bundle bundle = new Bundle();
41        bundle.putString("audioPath", AUDIO_PATH);   ← 绑定键-值对,把音乐文件
42        intent.putExtras(bundle);                       的路径传递给后台服务
43        startService(intent);   ← 启动服务
44        break;
45      case R.id.btnStop:   ← 资源中停止按钮的 id
46        if(intent != null)
```

```
47        {
48            stopService(intent);    ← 停止服务
49        }
50     break;
51    }
52  }
53 }
```

(2) 设计后台服务程序 AudioService.java。

```
1  package com.ex5_3;

2  import android.app.Service;
3  import android.content.Intent;
4  import android.media.MediaPlayer;
5  import android.os.Bundle;
6  import android.os.IBinder;
7  import android.util.Log;

9  public class AudioService extends Service
10 {
11   private MediaPlayer mediaPlayer = null;
12   private Intent intent2 = null;
13   private Bundle bundle2 = null;
14   private String audioPath ;
15   @Override
16   public IBinder onBind(Intent intent)
17   {
18     return null;
19   }
20   public int onStartCommand(Intent intent, int flags, int startId)
21   {
22     super.onStartCommand(intent, flags, startId);
23     audioPath = intent.getExtras().getString("audioPath");
24     /** 1.正在播放
25      *   使其暂停播放,并通知界面将Button的值改为"播放"
26      *   (如果正在播放,Button的值是"暂停")
27      */
28     if(mediaPlayer != null && mediaPlayer.isPlaying())
29     {
30         mediaPlayer.pause();
31         sendUpdateUI(1);//更新界面
32     }
33     /**
34      *2.正在暂停
35      */
```

```
36        else{
37            if(mediaPlayer == null)
38            {
39                mediaPlayer = new MediaPlayer();//如果被停止了，则为null
40                try {
41                    mediaPlayer.reset();
42                    mediaPlayer.setDataSource(audioPath) ;
43                    mediaPlayer.prepare();
44                } catch (Exception e)
45                { Log.e("player", "player prepare() err");     }
46            }
47            mediaPlayer.start();
48            sendUpdateUI(0);//更新界面
49        }
50     return START_STICKY;
51 }
52 @Override
53 public void onDestroy()
54 {
55     if(mediaPlayer !=null)
56     {
57         mediaPlayer.release();//停止时要release
58         sendUpdateUI(2);//更新界面
59     }
60     super.onDestroy();
61 }
62 private void sendUpdateUI(int flag)
63 {
64     intent2 = new Intent();    //action属性值
65     intent2.setAction("music");
66     bundle2 = new Bundle();
67     bundle2.putInt("backFlag", flag);//把flag传回去
68     intent2.putExtras(bundle2);
69     sendBroadcast(intent2);
70     /* 发送广播
71      * 后台服务把键名为backFlag的信息广播出去
72      * 发送后，Activity里的updateUIReceiver的onReceiver()方法
73      * 就能做相应的更新界面的工作了
74      */
75 }
76 }
```

(3) 设计广播接收器 Broadcast。

```
1 package com.ex5_3;
2 import android.content.BroadcastReceiver;
3 import android.content.Context;
4 import android.content.Intent;
```

```
5   /* 广播接收器 */
6   class Broadcast extends BroadcastReceiver
7   {
8     @Override
9     public void onReceive(Context context, Intent intent)
10    {
11      /**
12       * 更新界面，这里改变Button的值
13       * 从intent获取接收到的广播数据，
14       * 数据值为0表示此时是播放、为1表示是暂停、为2表示是停止
15       */
16      int backFlag = intent.getExtras().getInt("backFlag");
17      switch(backFlag)
18      {
19        case 0:
20          MainActivity.btnStart.setText("暂停");
21          break;
22        case 1:
23        case 2:
24          MainActivity.btnStart.setText("播放");
25          break;
26      }
27    }
28  }
```

（4）在配置文件AndroidManifest.xml中注册服务和广播。

```
1   <?xml version="1.0" encoding="utf-8"?>
2   <manifest xmlns:android="http://schemas.android.com/apk/res/android"
3       package="com.ex5_3"
4       android:versionCode="1"
5       android:versionName="1.0" >
6       <uses-sdk android:minSdkVersion="15" />
7       <application
8           android:icon="@drawable/ic_launcher"
9           android:label="@string/app_name" >
10          <activity
11              android:name=".MainActivity"
12              android:label="@string/app_name" >
13              <intent-filter>
14                  <action android:name="android.intent.action.MAIN" />
15                  <category android:name="android.intent.category.LAUNCHER" />
16              </intent-filter>
17          </activity>
18          <service android:enabled="true" android:name=".AudioService" />
19              <intent-filter>
20                  <action android:name="music" />
```

```
21              </intent-filter>
22          </application>
23      </manifest>
```

程序的运行结果如图 5.4 所示。

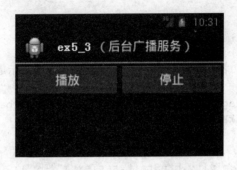

图 5.4 运行后台广播服务

5.3 系统服务

在第 4 章中学习了应用 Intent 进行参数传递，Android 系统提供了很多标准的系统服务，这些系统服务都可以很简单地通过 Intent 进行广播。

5.3.1 Android 的系统服务

Android 系统提供了大量的标准系统服务，这些系统服务用于完成不同的功能，Android 的系统服务如表 5-2 所示。

表 5-2 Android 的系统服务

系统服务	作用
WINDOW_SERVICE ("window")	窗体管理服务
LAYOUT_INFLATER_SERVICE ("layout_inflater")	布局管理服务
ACTIVITY_SERVICE ("activity")	Activity 管理服务
POWER_SERVICE ("power")	电源管理服务
ALARM_SERVICE ("alarm")	时钟管理服务
NOTIFICATION_SERVICE ("notification")	通知管理服务
KEYGUARD_SERVICE ("keyguard")	键盘锁服务
LOCATION_SERVICE ("location")	基于地图的位置服务
SEARCH_SERVICE ("search")	搜索服务
VIBRATOR_SERVICE ("vibrator")	振动管理服务
CONNECTIVITY_SERVICE ("connection")	网络连接服务
WIFI_SERVICE ("wifi")	Wi-Fi 连接服务
INPUT_METHOD_SERVICE ("input_method")	输入法管理服务
TELEPHONY_SERVICE ("telephony")	电话服务
DOWNLOAD_SERVICE ("download")	HTTP 协议的下载服务

系统服务实际上可以看作是一个对象，通过 Activity 类的 getSystemService 方法可以获得指定的对象（系统服务）。下面详细讲解几个常见的系统服务。

5.3.2 系统通知服务 Notification

Notification 是 Android 系统的一种通知服务，当手机来电、来短信、闹钟铃声时在状态栏中显示通知的图标和文字，提示用户处理，当拖动状态栏时可以查看这些信息。

Notification 提供了声音、振动等属性，表 5-3 列出了 Notification 的部分常见属性。

表 5-3 Notification 的部分常见属性

属性	说明
audioStreamType	所用的音频流的类型
contentIntent	设置单击通知条目所执行的 Intent
contentView	设置状态栏显示通知的视图
defaults	设置成默认值
deleteIntent	删除通知所执行的 Intent
icon	设置状态栏上显示的图标
iconLevel	设置状态栏上显示图标的级别
ledARGB	设置 LED 灯的颜色
ledOffMS	设置关闭 LED 时的闪烁时间（以毫秒计算）
ledOnMS	设置开启 LED 时的闪烁时间（以毫秒计算）
sound	设置通知的声音文件
tickerText	设置状态栏上显示的通知内容
vibrate	设置振动模式
when	设置通知发生的时间

系统通知服务 Notification 由系统通知管理对象 NotificationManager 进行管理及发布通知，由 getSystemService(NOTIFICATION_SERVICE)创建 NotificationManager 对象：

```
NotificationManager n_Manager =
    (NotificationManager)getSystemService(NOTIFICATION_SERVICE);
```

NotificationManager 对象通过 notify(int id, Notification notification) 方法把通知发送到状态栏，通过 cancelAll()方法取消以前显示的所有通知。

【例 5-4】 在状态栏中显示系统通知服务的应用示例。

在界面布局文件中设置两个按钮，分别为"发送系统通知"和"删除通知"。设计控制文件 MainActivity.java 如下：

```
1    package com.example.ex5_4;
2    import android.os.Bundle;
3    import android.app.Activity;
4    import android.app.Notification;
5    import android.app.NotificationManager;
6    import android.app.PendingIntent;
7    import android.content.Intent;
8    import android.view.View;
```

```
 9  import android.view.View.OnClickListener;
10  import android.widget.Button;
11
12  public class MainActivity extends Activity
13  {
14    NotificationManager n_Manager;
15    Notification notification;
16    Button btn1, btn2;
17    @Override
18    public void onCreate(Bundle savedInstanceState)
19    {
20        super.onCreate(savedInstanceState);
21        setContentView(R.layout.activity_main);
22        String service = NOTIFICATION_SERVICE;
23        n_Manager= (NotificationManager)getSystemService(service);
24        btn1=(Button)findViewById(R.id.btn1);
25        btn1.setOnClickListener(new mClick());
26        btn2=(Button)findViewById(R.id.btn2);
27        btn2.setOnClickListener(new mClick());
28    }
29    class mClick implements OnClickListener
30    {
31       @Override
32      public void onClick(View arg0)
33      {
34       if(arg0==btn1)
35       {
36          int icon = R.drawable.ic_con;
37          CharSequence tickerText = "紧急通知,程序已启动";
38          long when = System.currentTimeMillis();
39          notification=new Notification(icon, tickerText, when);
40          Intent intent = new Intent(MainActivity.this, MainActivity.class);
41          PendingIntent pi = PendingIntent.getActivity(MainActivity.this,
42                        0 , intent , 0);
43          notification.setLatestEventInfo(MainActivity.this,
44                        "通知", "通知内容", pi);
45          n_Manager.notify(0, notification);
46        }
47        else if(arg0==btn2)
48        {
49           n_Manager.cancelAll();
50        }
51     }
```

```
52    }
53 }
```

程序的运行结果如图 5.5 所示。

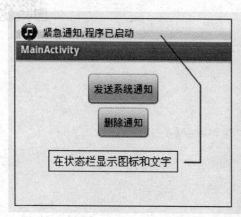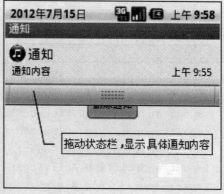

图 5.5　在状态栏中显示系统通知

5.3.3　系统定时服务 AlarmManager

系统定时服务 AlarmManager 又称为系统闹钟服务，其作用是在到达设定的时间后 AlarmManager 广播一个 Intent 信息。AlarmManager 常用的属性和方法见表 5-4。

表 5-4　AlarmManager 常用的属性和方法

属性或方法名称	说明
ELAPSED_REALTIME	设置闹钟时间，从系统启动开始
ELAPSED_REALTIME_WAKEUP	设置闹钟时间，从系统启动开始，如果设备休眠则唤醒
INTERVAL_DAY	设置闹钟时间，间隔一天
INTERVAL_FIFTEEN_MINUTES	间隔 15 分钟
INTERVAL_HALF_DAY	间隔半天
INTERVAL_HALF_HOUR	间隔半小时
INTERVAL_HOUR	间隔一小时
RTC	设置闹钟时间，从系统当前时间开始（System.currentTimeMillis()）
RTC_WAKEUP	设置闹钟时间，从系统当前时间开始，如果设备休眠则唤醒
set(int type,long tiggerAtTime, PendingIntent operation)	设置在某个时间执行闹钟
setRepeating(int type,long triggerAtTiem, long interval,PendingIntent operation)	设置在某个时间重复执行闹钟
setInexactRepeating(int type,long triggerAtTiem,long interval,PendingIntent operation)	在某个时间重复执行闹钟，但不是间隔固定时间
cancel(PendingIntent)	取消闹钟

AlarmManager 服务主要有下面两种应用：

(1) 在指定时长后执行某项操作;
(2) 周期性地执行某项操作。

下面通过示例说明这两种应用。

【例 5-5】 AlarmManager 时钟服务示例。

```
1   package com.example.ex5_5;
2   import java.util.Calendar;
3   import android.os.Bundle;
4   import android.os.SystemClock;
5   import android.view.View;
6   import android.view.View.OnClickListener;
7   import android.widget.Button;
8   import android.widget.Toast;
9   import android.app.Activity;
10  import android.app.AlarmManager;
11  import android.app.PendingIntent;
12  import android.content.Intent;
13
14  public class MainActivity extends Activity
15  {
16    Button btn1, btn2, btn3;
17    Intent intent;
18    PendingIntent sender;
19    @Override
20    public void onCreate(Bundle savedInstanceState)
21    {
22        super.onCreate(savedInstanceState);
23        setContentView(R.layout.activity_main);
24        btn1=(Button)findViewById(R.id.button1);
25        btn1.setOnClickListener(new mClick());
26        btn2=(Button)findViewById(R.id.button2);
27        btn2.setOnClickListener(new mClick());
28        btn3=(Button)findViewById(R.id.button3);
29        btn3.setOnClickListener(new mClick());
30    }
31    class mClick implements OnClickListener
32    {
33      @Override
34      public void onClick(View v)
35      {
36          switch (v.getId())
37          {
38              case R.id.button1:
39                  timing(); break;
40              case R.id.button2:
```

```
41                cycle(); break;
42            case R.id.button3:
43                cancel(); break;
44        }
45    }
46 }
47 /**
48  * 定时：5秒后发送一个广播，广播接收后Toast提示定时操作完成
49  */
50 void timing()
51 {
52     intent = new Intent(MainActivity.this, alarmreceiver.class);
53     intent.setAction("aaa");
54     sender = PendingIntent.getBroadcast(MainActivity.this, 0, intent, 0);
55     Calendar calendar = Calendar.getInstance();
56     calendar.setTimeInMillis(System.currentTimeMillis());
57     calendar.add(Calendar.SECOND, 5);        //设定一个5秒后的时间
58     AlarmManager alarm=(AlarmManager)getSystemService(ALARM_SERVICE);
59     alarm.set(AlarmManager.RTC_WAKEUP, calendar.getTimeInMillis(),
60               sender);
61     Toast.makeText(MainActivity.this, "5秒后alarm开启",
62               Toast.LENGTH_LONG).show();
63 }
64 /**
65  * 定义循环：每5秒发送一个广播，广播接收后Toast提示定时操作完成
66  */
67 void cycle()
68 {
69     Intent intent =new Intent(MainActivity.this, alarmreceiver.class);
70     intent.setAction("repeating");
71     PendingIntent sender = PendingIntent.getBroadcast(MainActivity.this,
72                    0, intent, 0);
73     /* 开始时间 */
74     long firstime=SystemClock.elapsedRealtime();
75     AlarmManager am=(AlarmManager)getSystemService(ALARM_SERVICE);
76     /* 5秒一个周期，不停地发送广播 */
77     am.setRepeating(AlarmManager.ELAPSED_REALTIME_WAKEUP ,
78                    firstime, 5*1000, sender);
79 }
80 /**
81  * 取消周期发送信息
82  */
83 void cancel()
84 {
```

```
85        Intent intent =new Intent(MainActivity.this, alarmreceiver.class);
86        intent.setAction("repeating");
87        PendingIntent sender=PendingIntent
88                .getBroadcast(MainActivity.this, 0, intent, 0);
89        AlarmManager alarm=(AlarmManager)getSystemService(ALARM_SERVICE);
90        alarm.cancel(sender);
91     }
92 }
```

程序的运行结果如图 5.6 所示。

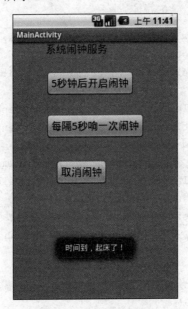

图 5.6 时钟服务示例

5.3.4 系统功能的调用

Android 系统通过 Intent 的 action 属性可以调用系统功能，常用的系统功能及调用语句见表 5-5。

表 5-5 常用的系统功能调用语句

系统功能	调用语句
浏览网页	Uri uri =Uri.parse("http://www.google.com");
	Intent it = new Intent(Intent.ACTION_VIEW,uri);
	startActivity(it);
从 Google 搜索内容	Intent intent = new Intent();
	intent.setAction(Intent.ACTION_WEB_SEARCH);
	intent.putExtra(SearchManager.QUERY,"searchString")
	startActivity(intent);
显示地图	Uri uri = Uri.parse("geo:38.899533,-77.036476");
	Intent it = new Intent(Intent.Action_VIEW,uri);
	startActivity(it);

续表

系统功能	调用语句
路径规划	Uri uri =Uri.parse("http://maps.google.com/maps?f=dsaddr=startLat%20start Lng&daddr=endLat%20endLng&hl=en"); Intent it = new Intent(Intent.ACTION_VIEW,URI); startActivity(it);
拨打电话	Uri uri =Uri.parse("tel:xxxxxx"); Intent it = new Intent(Intent.ACTION_DIAL,uri); startActivity(it);

发送短信程序	Intent it = new Intent(Intent.ACTION_VIEW);
	it.putExtra("sms_body", "TheSMS text");
	it.setType("vnd.android-dir/mms-sms");
	startActivity(it);

发送短信	Uri uri =Uri.parse("smsto:0800000123");
	Intent it = new Intent(Intent.ACTION_SENDTO, uri);
	it.putExtra("sms_body", "TheSMS text");
	startActivity(it);
	
	String body="this is sms demo";
	Intent mmsintent = new Intent(Intent.ACTION_SENDTO,
	  Uri.fromParts("smsto", number, null));
	mmsintent.putExtra(Messaging.KEY_ACTION_SENDTO_MESSAGE_BODY,body);
	mmsintent.putExtra(Messaging.KEY_ACTION_SENDTO_COMPOSE_MODE,true);
	mmsintent.putExtra(Messaging.KEY_ACTION_SENDTO_EXIT_ON_SENT,true);
	startActivity(mmsintent);

发送 Email	Uri uri =Uri.parse("mailto:xxx@abc.com");
	Intent it = new Intent(Intent.ACTION_SENDTO, uri);
	startActivity(it);
	Intent it = new Intent(Intent.ACTION_SEND);
	it.putExtra(Intent.EXTRA_EMAIL,"me@abc.com");
	it.putExtra(Intent.EXTRA_TEXT, "Theemail body text");
	it.setType("text/plain");
	startActivity(Intent.createChooser(it,"Choose Email Client"));
	
	Intent it=new Intent(Intent.ACTION_SEND);
	String[] tos={"me@abc.com"};
	String[]ccs={"you@abc.com"};
	it.putExtra(Intent.EXTRA_EMAIL, tos);
	it.putExtra(Intent.EXTRA_CC, ccs);
	it.putExtra(Intent.EXTRA_TEXT, "Theemail body text");
	it.putExtra(Intent.EXTRA_SUBJECT, "Theemail subject text");
	it.setType("message/rfc822");
	startActivity(Intent.createChooser(it,"Choose Email Client"));

发送邮件的附件	Intent it = new Intent(Intent.ACTION_SEND);
	it.putExtra(Intent.EXTRA_SUBJECT, "Theemail subject text");
	it.putExtra(Intent.EXTRA_STREAM,"file:///sdcard/mysong.mp3");
	sendIntent.setType("audio/mp3");
	startActivity(Intent.createChooser(it,"Choose Email Client"));

系统功能	调用语句

播放多媒体	Intent it = new Intent(Intent.ACTION_VIEW);
	Uri uri =Uri.parse("file:///sdcard/song.mp3");
	it.setDataAndType(uri,"audio/mp3");
	startActivity(it);
	Uri uri =Uri.withAppendedPath(MediaStore.Audio.Media.INTERNAL_CONTENT_URI,"1");
	Intent it = new Intent(Intent.ACTION_VIEW,uri);
	startActivity(it);

打开录音机	Intent mi = new Intent(Media.RECORD_SOUND_ACTION); startActivity(mi);

【例 5-6】 调用系统的短信发送功能示例。

本示例仅设置一个按钮，在按钮的事件中添加发送短信的代码。

（1）编写调用系统短信发送功能的源程序。

```
1    package com.example.ex5_6;
2    import android.net.Uri;
3    import android.os.Bundle;
4    import android.app.Activity;
5    import android.content.Intent;
6    import android.view.View;
7    import android.view.View.OnClickListener;
8    import android.widget.Button;
9
10   public class MainActivity extends Activity
11   {
12     Button btn_sms;
13     @Override
14     public void onCreate(Bundle savedInstanceState)
15     {
16       super.onCreate(savedInstanceState);
17       setContentView(R.layout.activity_main);
18       btn_sms=(Button)findViewById(R.id.btn1);
19       btn_sms.setOnClickListener(new mClick());
20     }
21     class mClick implements OnClickListener
22     {
23       @Override
24       public void onClick(View arg0)
25       {
26         Uri uri =Uri.parse("smsto:13900100100");
27         Intent it = new Intent(Intent.ACTION_SENDTO,uri);
28         it.putExtra("sms_body", "TheSMS text");
29         startActivity(it);
```

```
30      }
31   }
32 }
```

（2）在配置文件 AndroidManifest.xml 中添加访问网络权限的语句。

```
<uses-permission
        android:name="android.permission.INTERNET">
</uses-permission>
```

程序的运行结果如图 5.7 所示。

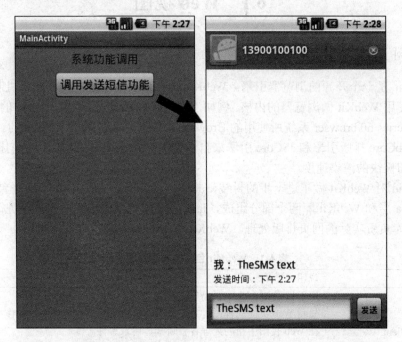

图 5.7　点击"调用发送短信功能"按钮调用系统的发送短信功能

习　题　5

1. 结合例 5-1 和例 5-3 编写一个具有较完善功能的后台音乐播放器。
2. 编写一个短信服务平台。

第 6 章　网络通信技术

6.1　Web 视图

6.1.1　浏览器引擎 WebKit

WebKit 是一个开源的浏览器引擎。WebKit 内核具有非常好的网页解析机制，很多应用系统都使用 WebKit 做浏览器的内核。例如，Google 的 Android 系统、Apple 的 iOS 系统、Nokia 的 Series 60 browser 系统所使用的 Browser 内核引擎都是基于 WebKit 的。WebKit 所包含的 WebCore 排版引擎和 JSCore 引擎来自于 KDE 的 KHTML 和 KJS，它们拥有清晰的源码结构和极快的渲染速度。

Android 对 WebKit 做了进一步的封装，并提供了丰富的 API。Android 平台的 WebKit 模块由 Java 层和 WebKit 库两个部分组成，Java 层负责与 Android 应用程序进行通信，而 WebKit 类库负责实际的网页排版处理。WebKit 包中的几个重要类见表 6-1。

表 6-1　WebKit 包中的几个重要类

类名	说明
WebSettings	用于设置 WebView 的特征、属性等
WebView	显示 Web 页面的视图对象，用于网页数据的载入、显示等操作
WebViewClient	在 Web 视图中帮助处理各种通知、请求事件
WebChromeClient	Google 浏览器 Chrome 的基类，辅助 WebView 处理 JavaScript 对话框、网站的标题、网站的图标、加载进度条等

6.1.2　Web 视图对象

1. WebView 类

在 WebKit 的 API 包中最重要、最常用的类是 Android.webKit.WebView。WebView 类是 WebKit 模块 Java 层的视图类，所有需要使用 Web 浏览功能的 Android 应用程序都要创建该视图对象，用于显示和处理请求的网络资源。目前，WebKit 模块支持 HTTP、HTTPS、FTP 以及 JavaScript 请求。WebView 作为应用程序的 UI 接口为用户提供了一系列的网页浏览、用户交互接口，客户程序通过这些接口访问 WebKit 核心代码。

WebView 类的常用方法见表 6-2。

表 6-2　WebView 类的常用方法

方法	说明
WebView（Context context）	构造方法

续表

方法	说明
loadUrl（String url）	加载 URL 网站页面
loadData（String data, String mimeType, String encod）	显示 HTML 格式的 Web 视图
reload()	重新加载网页
getSettings()	获取 WebSettings 对象
goBack()	返回上一页面
goForward()	向前一页面
clearHistory()	清除历史记录
addJavascriptInterface （Object obj, String interfaceName）	将对象绑定到 JavaScript，允许从网页控制 Android 程序，从网页调用该对象的方法

2. 使用 WebView 的说明

（1）设置 WebView 的基本信息。

- 如果访问的页面中有 JavaScript，则 WebView 必须设置支持 JavaScript：

```
webview.getSettings().setJavaScriptEnabled(true);
```

- 触摸焦点起作用：

```
requestFocus();
```

- 取消滚动条：

```
this.setScrollBarStyle(SCROLLBARS_OUTSIDE_OVERLAY);
```

（2）设置 WebView 要显示的网页。

- 互联网用 webView.loadUrl("http://www.google.com");
- 本地文件用 webView.loadUrl("file:///android_asset/XX.html")，本地文件要存放在项目的 assets 目录中。

（3）用 WebView 点击链接看了很多页面以后，如果不做任何处理，按 Back 键浏览器会调用 finish()结束自身的运行；如果希望浏览的网页回退而不是退出浏览器，需要在当前 Activity 中覆盖 Activity 类的 onKeyDown(int keyCoder,KeyEvent event)方法处理该 Back 事件。

```
public boolean onKeyDown(int keyCoder,KeyEvent event)
{
    if(webView.canGoBack() && keyCoder == KeyEvent.KEYCODE_BACK)
    {
        webview.goBack();    //goBack()表示返回WebView的上一页面
        return true;
    }
    return false;
}
```

【例 6-1】 应用 WebView 对象浏览网页。

(1)设计界面布局文件 activity_main.xml。在界面布局中设置了一个文本编辑框,用于输入网址;设置了一个按钮,用于打开网页;还设置了一个网页视图组件 WebView,用于显示网页。

```xml
1   <?xml version="1.0" encoding="utf-8"?>
2   <LinearLayout xmlns:android="http://schemas.android.com/apk/res/android"
3       android:layout_width="fill_parent"
4       android:layout_height="fill_parent"
5       android:layout_gravity="center_horizontal"
6       android:orientation="vertical" >
7       <LinearLayout
8           android:id="@+id/LinearLayout2"
9           android:layout_width="fill_parent"
10          android:layout_height="wrap_content" >
11          <EditText
12              android:id="@+id/editText1"
13              android:layout_width="207dp"
14              android:layout_height="wrap_content"/>
15          <Button
16              android:id="@+id/button1"
17              android:layout_width="wrap_content"
18              android:layout_height="wrap_content"
19              android:layout_weight="1"
20              android:text="打开网页" />
21      </LinearLayout>
22      <WebView
23          android:id="@+id/webView1"
24          android:layout_width="fill_parent"
25          android:layout_height="fill_parent" />
26  </LinearLayout>
```

(2)设计控制文件 MainActivity.java。

```java
1   package com.example.ex6_1;
2   import android.os.Bundle;
3   import android.app.Activity;
4   import android.view.View;
5   import android.view.View.OnClickListener;
6   import android.webkit.WebView;
7   import android.widget.Button;
8   import android.widget.EditText;
9
10  public class MainActivity extends Activity
11  {
12      WebView webView;
```

```
13    Button openWebBtn;
14    EditText edit;
15    @Override
16    public void onCreate(Bundle savedInstanceState)
17    {
18        super.onCreate(savedInstanceState);
19        setContentView(R.layout.activity_main);
20        openWebBtn = (Button)findViewById(R.id.button1);
21        edit =(EditText)findViewById(R.id.editText1);
22        openWebBtn.setOnClickListener(new mClick());
23    }
24    class mClick implements OnClickListener
25    {
26       public void onClick(View arg0)
27       {
28           String url = edit.getText().toString();
29           webView = (WebView)findViewById(R.id.webView1);
30           webView.loadUrl("http://" + url);
31       }
32    }
33 }
```

（3）在配置文件中加入网络权限。网络程序需要在配置文件 AndroidManifest.xml 中加入允许访问网络的权限语句：

```
<uses-permission android:name="android.permission.INTERNET" />
```

添加权限后的 AndroidManifest.xml 程序如下：

```
1  <?xml version="1.0" encoding="utf-8"?>
2  <manifest xmlns:android="http://schemas.android.com/apk/res/android"
3          package="com.example.ex6_1">
4    <application
5        android:allowBackup="true"
6        android:icon="@mipmap/ic_launcher"
7        android:label="@string/app_name"
8        android:supportsRtl="true"
9        android:theme="@style/AppTheme">
10      <activity android:name=".MainActivity">
11         <intent-filter>
12             <action android:name="android.intent.action.MAIN" />
13             <category android:name="android.intent.category.LAUNCHER" />
14         </intent-filter>
15      </activity>
16    </application>
```

```
17    <uses-permission android:name="android.permission.INTERNET" />
18 </manifest>
```

程序的运行结果如图 6.1 所示。

图 6.1　用 WebView 显示网页

6.1.3　调用 JavaScript

1. 几个辅助类

1）WebSettings 类

WebView 对象刚创建时使用的是系统的默认设置,当需要对 WebView 对象的属性等做自定义设置时需要用到 WebSettings 类。WebSettings 类的常用方法见表 6-3。

表 6-3　WebSettings 类的常用方法

方法	说明
setAllowFileAccess(boolean flag)	设置是否允许访问文件数据
setJavaScriptEnabled(boolean flag)	设置是否支持 JavaScript 脚本
setBuiltInZoomControls(boolean flag)	设置是否支持缩放
setBlockNetworkImage(boolean flag)	设置是否禁止显示图片,true 为禁止显示
setDefaultFontSize(int size)	设置默认字体大小,在 1～72 取值
setTextZoom(int textZoom)	设置页面文字缩放的百分比,默认为 100

2）WebViewClient 类

WebViewClient 类用于对 WebView 对象中各种事件的处理,通过重写提供的这些事件方法可以对 WebView 对象在页面载入、资源载入、页面访问错误等情况发生时进行各种操作。WebViewClient 类的常用方法见表 6-4。

表 6-4　WebViewClient 类的常用方法

方法	说明
onLoadResource（WebView view, String url）	通知 WebView 加载 url 指定的资源时触发
onPageStarted（WebView view, String url, Bitmap favicon）	页面开始加载时触发
onPageFinished（WebView view, String url）	页面加载完毕时触发

3）WebChromeClient 类

WebChromeClient 是辅助 WebView 处理 JavaScript 对话框、网站的标题、网站的图标、加载进度条等操作的类，其常用方法见表 6-5。

表 6-5　WebChromeClient 类的常用方法

方法	说明
onJsAlert(WebView view, String url, String message, JsResult result)	处理 JavaScript 的 Alert 对话框
onJsPrompt(WebView view, String url, String message, String defaultValue, JsPromptResult result)	处理 JavaScript 的 Prompt 提示对话框
onCloseWindow(WebView window)	关闭 WebView

2. 调用本地 HTML 网页文件的 JavaScript

用户可以在 Android 程序中调用本地的 HTML 网页文件的 JavaScript，例如下面的例子。

【例 6-2】　在 Android 程序中调用本地的 HTML 程序示例。

（1）在 Android Studio 编辑器中首先调整成 Project 模式，然后在 main 目录下新建 assets 目录，在 assets 目录下新建一个 HTML 程序 test.html。通常，assets 目录存放应用程序所使用的外部资源文件，而 res 目录存放应用程序自身的资源文件，如图 6.2 所示。

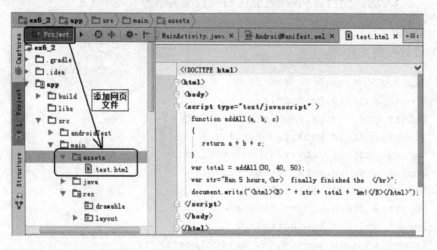

图 6.2　在 "assets" 目录下新建 test.html

HTML 程序 test.html 的代码如下：

```
1   <HTML>
2   <head>
```

```
3      <title> 一个简单的JavaScript示例  </title>
4    </head>
5    <body>
6    <script language="javascript" type="text/javascript" >
7       function addAll(a, b, c)
8       {
9          return a + b + c;
10      }
11      var total = addAll(30, 40, 50);
12      var str="Ran 5 hours,<br>  finally finished the  ";
13      document.write("<html><B> " + str + total + "  km!</B></html>");
14   </script>
15   </body>
16   </HTML>
```

（2）设计界面布局文件。

```
1   <RelativeLayout xmlns:android="http://schemas.android.com/apk/res/android"
2      xmlns:tools="http://schemas.android.com/tools"
3      android:layout_width="fill_parent"
4      android:layout_height="fill_parent" >
5      <WebView
6         android:id="@+id/webView1"
7         android:layout_width="fill_parent"
8         android:layout_height="fill_parent"  />
9   </RelativeLayout>
```

（3）设计控制文件。

```
1   package com.example.ex6_2;
2   import android.os.Bundle;
3   import android.os.Handler;
4   import android.webkit.JsResult;
5   import android.webkit.WebChromeClient;
6   import android.webkit.WebSettings;
7   import android.webkit.WebView;
8   import android.widget.Toast;
9   import android.app.Activity;
10  import android.webkit.javascriptInterface;
11  public class MainActivity extends Activity
12  {
13    WebView webView;
14    Handler handler = new Handler();
15    MWebChromeClient mWebChromeClient;
16    @Override
```

```
17  public void onCreate(Bundle savedInstanceState)
18  {
19    super.onCreate(savedInstanceState);
20    setContentView(R.layout.activity_main);
21    webView = (WebView) findViewById(R.id.webView1);
22    WebSettings webSettings = webView.getSettings();
23    webSettings.setAllowFileAccess(true);//设置允许访问文件数据
24    webSettings.setJavaScriptEnabled(true);//设置支持JavaScript脚本
25    webSettings.setBuiltInZoomControls(true);//设置支持缩放
26    webSettings.setDefaultFontSize (24);
27    MObject mObject = new MObject();
28    webView.addJavascriptInterface(mObject, "test");
29    mWebChromeClient = new MWebChromeClient();
30    webView.setWebChromeClient(mWebChromeClient);
31    webView.loadUrl("file:///android_asset/test1.html");
32  }
33  class MObject extends Object
34  { @Java scriptInterface
35    public void android_show()
36    {
37      handler.post(new Runnable()
38      {
39        public void run()
40        {
41          System.out.println("提示：调用了多线程的run()方法!!");
42          webView.loadUrl("javascript: show_alert()");   ◀── 调用 JavaScript 函数
43        }
44      });
45    }
46  }
47  class MWebChromeClient extends WebChromeClient
48  {
49    @Override
50    public boolean onJsAlert(WebView view,   ◀── 处理 JavaScript 的 Alert 对话框
51        String url, String message, JsResult result)
52    {
53      Toast.makeText(getApplicationContext(), message,
54          Toast.LENGTH_LONG).show();
55      return true;
56    }
57  }
58  }
```

程序的运行结果如图 6.3 所示。

图 6.3 调用 HTML 的 JavaScript

【例 6-3】 用 Android 程序操纵 JavaScript 对话框。

（1）在 Android Studio 编辑器中调整成 Project 模式，然后在 main 目录下新建 assets 目录，在 assets 目录下新建一个 JavaScript 对话框程序 test1.html，其代码如下：

```
1   <html>
2   <head>
3     <title>JavaScript与Android交互</title>
4   </head>
5   <script type="text/javascript">
6    function show_alert()
7    {
8      var a = document.getElementById("text").value;
9      alert("Hello " + a );
10   }
11  </script>
12  <body>
13  <form action="">
14    <input type="text" id="text" value=""/>
15    <input type="button" id="button"
16          onclick="window.test.android_show()"
17          value="call Android"/>
18  </form>
19  </body>
20  </html>
```

第 8、9 行：JavaScript 对话框

第 15～17 行：调用 Android 程序标记为 test 的实例对象的函数

（2）设计界面布局文件。

```
1   <RelativeLayout xmlns:android="http://schemas.android.com/apk/res/
```

```
          android"
2      xmlns:tools="http://schemas.android.com/tools"
3      android:layout_width="fill_parent"
4      android:layout_height="fill_parent" >
5      <WebView
6          android:id="@+id/webView1"
7          android:layout_width="fill_parent"
8          android:layout_height="fill_parent"  />
9  </RelativeLayout>
```

（3）设计控制文件。

```
1   package com.example.ex6_3;
2   import android.os.Bundle;
3   import android.os.Handler;
4   import android.webkit.JsResult;
5   import android.webkit.WebChromeClient;
6   import android.webkit.WebSettings;
7   import android.webkit.WebView;
8   import android.widget.Toast;
9   import android.app.Activity;
10
11  public class MainActivity extends Activity
12  {
13    WebView webView;
14    Handler handler = new Handler();
15    MWebChromeClient mWebChromeClient;
16    @Override
17    public void onCreate(Bundle savedInstanceState)
18    {
19      super.onCreate(savedInstanceState);
20      setContentView(R.layout.activity_main);
21      webView = (WebView) findViewById(R.id.webView1);
22      WebSettings webSettings = webView.getSettings();
23      webSettings.setAllowFileAccess(true);      //设置允许访问文件数据
24      webSettings.setJavaScriptEnabled(true); //设置支持JavaScript脚本
25      webSettings.setBuiltInZoomControls(true);     //设置支持缩放
26      webSettings.setDefaultFontSize (24);
27      MObject mObject = new MObject();
28      webView.addJavascriptInterface(mObject, "test");
29      mWebChromeClient = new MWebChromeClient();
30      webView.setWebChromeClient(mWebChromeClient);
31      webView.loadUrl("file:///android_asset/test1.html");
32    }
33    class MObject extends Object{
34      @JavascriptInterface     ← 为了安全问题，"@JavascriptInterface"注解不可少
```

```
35    public void android_show()
36    {
37     handler.post(new Runnable()
38     {
39      public void run()
40      {
41       System.out.println("提示：调用了多线程的run()方法!!");
42       webView.loadUrl("javascript: show_alert()");     ← 调用 JavaScript 函数
43      }
44     });
45    }
46   }
47   class MWebChromeClient extends WebChromeClient
48   {
49    @Override
50    public boolean onJsAlert(WebView view,     ← 处理 JavaScript 的 Alert 对话框
51           String url, String message, JsResult result)
52    {
53     Toast.makeText(getApplicationContext(), message,
54          Toast.LENGTH_LONG).show();
55     return true;
56    }
57   }
58  }
```

程序的运行结果如图 6.4 所示。

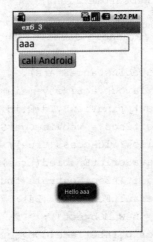

图 6.4　操纵 JavaScript 对话框

6.2　基于 TCP 协议的网络程序设计

本节将介绍使用 Android 编写网络通信程序的一些实例，其中重点介绍客户机/服务器

的应用程序及 Web 视图应用程序的设计方法。

6.2.1 网络编程的基础知识

1. IP 地址

在网络中连接了很多计算机,假设计算机 A 向计算机 B 发送信息,若网络中还有第 3 台计算机 C,那么主机 A 怎么知道信息被正确地传送到主机 B 而不是被传送到主机 C 中了呢?如图 6.5 所示。

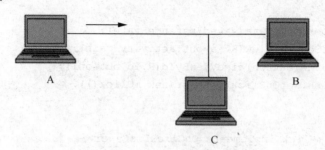

图 6.5 主机 A 向主机 B 发送信息

网络中的每台计算机都必须有一个唯一的 IP 地址作为标识,这个数通常写作一组由"."号分隔的十进制数,例如思维论坛的服务器地址为 218.5.77.187。正如大家所见 IP 地址均由 4 个部分组成,每个部分的范围都是 0~255,以表示 8 位地址。

值得注意的是 IP 地址都是 32 位地址,这是 IP 协议版本 4(简称 IPv4)规定的,目前由于 IPv4 地址已近耗尽,所以 IPv6 地址正逐渐代替 IPv4 地址,IPv6 地址是 128 位无符号整数。

在 Java.net 包中 IP 地址由一个被称为 InetAddress 的特殊类来描述。这个类提供了 3 个用来获得一个 InetAddress 类的实例的静态方法。这 3 个方法如下。

- getLocalHost():返回一个本地主机的 IP 地址。
- getByName(String host):返回对应于指定主机的 IP 地址。
- getAllByName(String host):对于某个主机有多个 IP 地址(多宿主机),可用于得到一个 IP 地址数组。

此外,对一个 InetAddress 的实例可以使用以下方法。

- getAddress():获得一个用字节数组形式表示的 IP 地址。
- getHostName():做反向查询,获得对应于某个 IP 地址的主机名。

【例 6-4】 通过域名查找 IP 地址。

```
1   package com.example.iptest;
2   import java.net.InetAddress;
3   import java.net.UnknownHostException;
4   import android.os.Bundle;
5   import android.view.View;
6   import android.view.View.OnClickListener;
7   import android.widget.Button;
```

```
8    import android.widget.Toast;
9    import android.app.Activity;
10
11   public class MainActivity extends Activity
12   {
13     Button IPBtn;
14     @Override
15     public void onCreate(Bundle savedInstanceState)
16     {
17         super.onCreate(savedInstanceState);
18         setContentView(R.layout.activity_main);
19         IPBtn=(Button)findViewById(R.id.button1);
20         IPBtn.setOnClickListener(new mClick());
21     }
22
23     class mClick implements OnClickListener
24     {
25       @Override
26       public void onClick(View arg0)
27       {
28         String str;
29         try{
30             InetAddress zsm_address=InetAddress.getByName("www.zsm8.com");
31           str="思维论坛的IP地址为：\n"+zsm_address.toString();
32         }
33         catch(UnknownHostException e)
34         {
35             str="无法找到思维论坛";
36         }
37         Toast.makeText(MainActivity.this, str, Toast.LENGTH_LONG).show();
38     }
39   }
40 }
```

网络程序需要在配置文件 AndroidManifest.xml 中加入允许访问网络的权限语句：

```
<uses-permission android:name="android.permission.INTERNET" />
```

程序运行结果如图 6.6 所示。

若在上面的例子中将第 30 行的 getByName()方法改为 getLocalHost()方法，则显示设备的 IP 地址。

2. 端口

由于一台计算机上可同时运行多个网络程序，IP 地址只能保证把数据信息送到该计算机，但无法知道要把这些数据交给该主机上的哪个网络程序，因此用"端口号"来标识正在计算机上运行的进程（程序）。每个被发送的网络数据包也都包含有"端口号"，用于将该数据帧交给具有相同端口号的应用程序来处理。

图 6.6 通过域名查找 IP 地址

例如在一个网络程序指定自己所用的端口号为 52000，那么其他网络程序（比如端口号为 13）发送给这个网络程序的数据包必须包含 52000 端口号，当数据到达计算机后驱动程序根据数据包中的端口号就知道要将这个数据包交给这个网络程序，如图 6.7 所示。

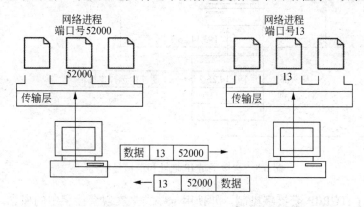

图 6.7 用"端口号"来标识进程

端口号是一个整数，其取值范围为 0～65 535。同一台计算机上不能同时运行两个有相同端口号的进程。通常 0～1023 的端口号作为保留端口号，用于一些网络系统服务和应用，用户的普通网络应用程序应该使用 1024 以后的端口号，从而避免端口号冲突。

3. TCP 与 UDP 协议

在网络协议中有两个高级协议是编写网络应用程序时常用的，它们是"传输控制协议"（Transmission Control Protocol，TCP）和"用户数据报协议"（User Datagram Protocol，UDP）。

TCP 是面向连接的通信协议，TCP 提供两台计算机之间的可靠、无差错的数据传输。在应用程序利用 TCP 进行通信时信息源与信息目标之间会建立一个虚连接。这个连接一旦建成，两台计算机之间就可以把数据当作一个双向字节流进行交换。接收方对于接收到的每一个数据包都会发送一个确认信息，发送方只有在收到接收方的确认信息后才发送下一个数据包，通过这种确认机制保证数据传输无差错。

UDP 是无连接通信协议，UDP 不保证可靠数据的传输。简单地说，如果一个主机向另外一台主机发送数据，这一数据就会立即发送，而不管另外一台主机是否已准备接收数据。如果另外一台主机收到了数据，它不会确认收到与否。这一过程类似于从邮局发送信件，我们无法确定收信人一定收到了发出去的信件。

4. 套接字

我们已经知道，通过 IP 地址可以在网络上找到主机，通过端口可以找到主机上正在运行的网络程序。在 TCP/IP 通信协议中，套接字（Socket）就是 IP 地址与端口号的组合。如图 6.8 所示，IP 地址 193.14.26.7 与端口号 13 组成一个套接字。

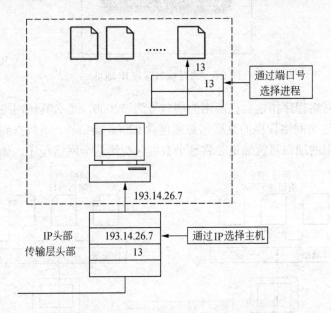

图 6.8　套接字是 IP 地址和端口号组合

Java 使用了 TCP/IP 套接字机制，并使用一些类来实现套接字中的概念。Java 中的套接字提供了在一台处理机上执行的应用程序与在另一台处理机上执行的应用程序之间进行连接的功能。

准确地说，网络通信不能仅说成是两台计算机之间在通信，而是两台计算机上执行的网络应用程序（进程）之间在收发数据。

当两个网络程序需要通信时它们可以通过使用 Socket 类建立套接字连接。我们可以把套接字连接想象为一个电话呼叫，当呼叫完成后通话的任何一方都可以随时讲话，但是在最初建立呼叫时必须有一方主动呼叫，而另一方正在监听铃声，我们把呼叫方称为"客户端"，把负责监听的一方称为"服务器端"。

5. 在客户端建立套接字 Socket 对象

在客户端使用 Socket 类建立向指定服务器 IP 和端口号连接的套接字，其构造方法如下：

```
Socket(host_IP, prot);
```

其中 host_IP 是服务器的 IP 地址，prot 是一个端口号。

由于建立 Socket 对象可能发生 IOException 异常，因此在建立 Socket 对象时要使用 try-cahch 结构处理异常事件。

Socket 的主要方法如下。

- getInputStream()：获得一个输入流，读取从网络线路上传送来的数据信息。
- getOutputStream()：获得一个输出流，用这个输出流将数据信息写入到网络"线路"上。

6. 在服务器端建立套接字 Socket 对象

在编写 TCP 网络服务器程序时首先要用 ServerSocket 类创建服务器 Socket，ServerSocket 类的构造方法如下：

```
ServerSocket(int port);
```

创建 ServerSocket 实例是不需要指定 IP 地址的，ServerSocket 总是处于监听本机端口的状态。

ServerSocket 类的主要方法如下：

```
Socket accept();
```

该方法用于在服务器端的指定端口监听客户机发起的连接请求，并与之连接，其返回值为 Socket 对象。

6.2.2 利用 Socket 设计客户机/服务器系统程序

基于 TCP 协议的网络程序都是采用客户机/服务器系统模式。利用套接字 Socket 设计客户机/服务器系统程序进行数据通信与传输大致有以下几个步骤：

（1）在计算机上创建服务器端 ServerSocket，设置建立连接的端口号。
（2）创建 Android 客户端 Socket 对象，设置绑定的主机名称或 IP 地址，指定连接端口号。
（3）客户机 Socket 发起连接请求。
（4）建立连接。
（5）取得 InputStream 和 OutputStream。
（6）利用 InputStream 和 OutputStream 进行数据传输。
（7）关闭 Socket 和 ServerSocket。

客户机/服务器模式的连接请求与响应过程如图 6.9 所示。

【例 6-5】 远程数据通信示例，本例由 Android 客户端程序和计算机服务器端程序两部分组成。

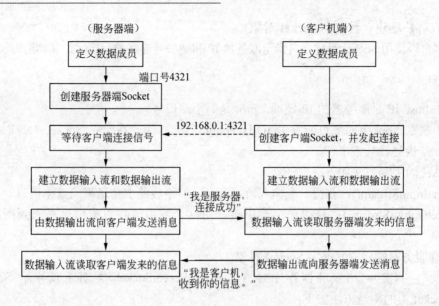

图 6.9 客户机/服务器模式

(1) Android 客户端程序。

```
1   package com.example.ex6_5;

2   import java.io.DataInputStream;
3   import java.io.DataOutputStream;
4   import java.net.Socket;
5   import android.os.StrictMode;
6   import android.support.v7.app.AppCompatActivity;
7   import android.os.Bundle;
8   import android.view.View;
9   import android.widget.Button;
10  import android.widget.TextView;
11  public class SocketClientActivity extends AppCompatActivity
12  {
13      private Socket socket=null;
14      private DataInputStream dis=null;
15      private DataOutputStream dos=null;
16      private TextView mTextView1;
17      private Button Button1;
18      String msg="";
19      @Override
20      public void onCreate(Bundle savedInstanceState)
21      {
22          super.onCreate(savedInstanceState);
23          setContentView(R.layout.main);
24          mTextView1 = (TextView)findViewById(R.id.textView);
25          Button1 = (Button) findViewById(R.id.Button);
```

```
26      Button1.setOnClickListener(new  mClick());
27      //以下代码避免程序出现NetworkOnMainThreadException异常
28      StrictMode.setThreadPolicy(new StrictMode
29              .ThreadPolicy
30              .Builder()
31              .detectDiskWrites()
32              .detectDiskReads()
33              .detectNetwork()
34              .penaltyLog()
35              .build() );
36    }
37    class mClick implements View.OnClickListener {
38     @Override
39     public void onClick(View v)
40     {
41         Client();
42      }
43     public void Client()
44     {
45    try {
46     socket = new Socket("192.168.0.1", 4321);
47        }catch (Exception ioe) {
48          System.out.print("socket  err ");
49        }
50    try{
51          //创建输入流对象dis读取数据，创建输出流对象dos发送数据
52          dis = new DataInputStream(socket.getInputStream());
53          dos = new DataOutputStream(socket.getOutputStream());
54         dos.writeUTF("给我数据啊。。");
55         dos.flush();
56        } catch (IOException ioe) {
57           System.out.print("DataStream create err ");
58        }
59        ReadStr();      ◀── 读取服务器发来的数据
60        try{
61           Thread.sleep(500);
62           msg = "手机客户端发来贺电！";
63           WriteString(msg);   ◀── 发送数据
64       dis.close();
65       socket.close();
66         }catch (Exception ioe
67    { System.out.println("socket close() err .......");  }
68         ReadStr();   ◀── 读取服务器发来的第2条数据
69    }
70          //写数据到socket，即发送数据
```

```
71    public void WriteString(String str)
72    {
73    try {
74            dos.writeUTF(str);         // 发送字符串 str 的数据
75            dos.flush();
76            ReadStr();
77            socket.close();
78         } catch (IOException e) {
79            System.out.print("WriteString() err");
80         }
81    }
82      //显示从socket返回的数据，即读取数据
83      public void ReadStr()
84    {
85      try {
86
87          if((msg = dis.readUTF()) != null)   // 读取数据存放到字符串变量 msg
88          {
89
90              mTextView1.append(msg);
91          }
92      } catch (IOException ioe) {
93          System.out.print("ReadStr() err ");
94      }
95    }
96  }
```

（2）服务器端程序。

```
1   import java.io.DataOutputStream;
2   import java.io.DataInputStream;
3   import java.io.IOException;
4   import java.net.ServerSocket;
5   import java.net.Socket;
6
7   public class server
8   {
9     private ServerSocket ss;
10    private Socket socket;
11    private DataInputStream dis;
12    private DataOutputStream dos;
13    public server()
14    {
15       new ServerThread().start();
16    }
17    class ServerThread extends Thread
```

```
18    {
19        public void run()
20        {
21            try {
22                ss=new ServerSocket(4321);          ← 实例化服务器端套接字对象
23                System.out.println("服务器启动了");
24                while(true)
25                {
26                    socket = ss.accept();            ← 阻塞端口,等待客户机连接
27                    System.out.println("有客户端连接到服务器");
28                    dis = new DataInputStream(socket.getInputStream());
29                    dos = new DataOutputStream(socket.getOutputStream());
30                    dos.writeUTF("恭喜你,连接服务器成功! \n");   ← 向手机发送一条数据
31                    dos.flush();
32                    System.out.println("服务器休眠20秒......");
33                Thread.sleep(500);
34                 String msg="";
35                if((msg = dis.readUTF()) != null) {
36                     System.out.println(msg);
37                 }
38                 dos.writeUTF("你发来的数据服务器收到了。^_^");  ← 发送第2条数据
39                 dos.flush();
40                 }
41            }
42                catch (Exception e) {System.out.println("读写错误");}
43                finally{
44                    try {
45                        dis.close();
46                        dos.close();
47                    } catch (IOException e) {e.printStackTrace();}
48                }
49            }
50    }
51    public static void main(String[] args) throws IOException
52    {
53        new server();
54    }
55 }
```

(3) 在配置文件 AndroidManifest.xml 中加入允许访问网络的权限语句:

```
<uses-permission android:name="android.permission.INTERNET" />
```

程序由客户机程序和服务器端程序两部分组成。
(1) 客户机程序:
① 程序的第 28~35 行为设置线程策略,避免程序出现 NetworkOnMainThreadException

异常，这是 Android 与 Java 不同的地方。

② 在第 46 行创建一个可以连接到 server 的套接字，其端口为 4321。在运行程序时，当程序执行到该语句立即向服务器发起连接。

③ 在第 59 行调用 ReadStr()方法，通过数据输入流读取从服务器发送到"线路"里的信息。

④ 在第 63 行调用 WriteString（String str）方法，通过数据输出流向由套接字建立的连接"线路"（向服务器端方向）发送信息。

（2）服务器端程序：

① 在第 17 行创建多线程，可以用于多客户端的连接。

② 在第 22 行创建服务器端套接字，设定其端口号为 4321，该端口号与客户机套接字的端口号必须一致。注意，这里要使用 try-catch 结构处理异常事件。

③ 在第 26 行服务器端套接字对象使用 accept()方法监听端口，等待接收客户机传来的连接信号。

④ 在第 28、29 行建立套接字的数据输入流 dis 及数据输出流 dos。

⑤ 第 30 行通过数据输出流向由套接字建立的连接"线路"（向客户机方向）发送"连接已经建立"的信息。

⑥ 在第 35 行通过数据输入流读取客户机发送在"线路"里的信息。

⑦ 在第 37 行显示接收到的信息。

将服务器端程序保存为 SServer.java，编译程序。首先运行服务器程序，然后启动模拟器运行客户端程序。

程序运行结果如图 6.10 所示（先运行服务器端程序，再运行客户端程序）。

(a) 服务器端运行结果　　　　　　(b) 客户端运行结果

图 6.10　远程数据传输

6.2.3　应用 Callable 接口实现多线程 Socket 编程

1. Callable 接口

在 Java 语言中，经常使用的 Thread 类在 run()方法执行完以后是没有返回值的，要实现子线程完成任务后返回值给主线程需要借助第三方转存。Callable 接口提供了一种有返回值的多线程实现方法。

Callable 接口的定义如下：

```
public interface Callable<V>
{
    V call() throws Exception;
}
```

2. 线程接口 Runnable 和 Callable 的区别

Callable 接口与 Java 的 Runnable 接口的作用很类似，但它们之间又有很多不同，其区别如下：

（1）Runnable 接口自从 Java 1.1 就有了，而 Callable 接口是在 1.5 之后新增加的。

（2）Callable 接口中定义的方法是 call()，Runnable 中定义的方法是 run()。

（3）Callable 接口的任务执行后可以有返回值，而 Runnable 接口的任务是不能有返回值的(其返回类型为 void)。

（4）call()方法可以抛出异常，run()方法不能抛出异常。

（5）运行 Callable 接口任务可以返回一个 Future 对象，该对象为异步计算的结果。它提供了检查计算是否完成的方法，以等待计算的完成，并检索计算的结果。用户通过 Future 对象可以了解任务的执行情况，可取消任务的执行，还可获取执行结果。

3. 应用 Callable 接口实现多线程 Socket 编程示例

【例 6-6】 应用 Callable 接口实现多线程远程数据通信示例。

Android 客户端程序由实现 Callable 接口的 connSocket.java 和主程序 MainActivity.java 两部分组成。

（1）实现 Callable 接口的 connSocket.java。

```
1   package  （略）
2   import   （略）
3   class connSocket implements Callable<String>
4   {
5       static DataInputStream datain;
6       static DataOutputStream dataout;
7       static Socket ss;
8       String IP="58.199.89.161";
9       private String runlog=" ";
10      public String call() throws Exception         ←定义 Callable 接口的 call()方法
11      {
12          try {
13              ss = new Socket();
14              SocketAddress socketAddress = new InetSocketAddress(IP, 8888);
15              ss.connect(socketAddress, 5000);                //设置超时时间
16              datain=new DataInputStream(ss.getInputStream());//创建数据输入流
```

```
17      dataout=new DataOutputStream(ss.getOutputStream());//创建数据输出流
18      dataout.writeUTF("客户端发来的信息: Socket 我来了 !! ");
19      this.runlog= datain.readUTF();
20      Thread.sleep(500);
21      dataout.writeUTF("客户端发来的信息: 我已经收到服务器的信息 !! ");
22      this.runlog= datain.readUTF();
23    }catch (Exception e){ this.runlog="Socket错误";}
24    return  this.runlog;
25  }

26  public static void disConnet()
27  {
28   if(datain != null)
29     try{ datain.close(); }catch (Exception e){ e.printStackTrace(); }
30   if(dataout != null)
31     try{ dataout.close();}catch (Exception e){ e.printStackTrace(); }
32   if(ss != null)
33     try{ ss.close();}catch (Exception e){ e.printStackTrace(); }
34  }
35 }
```

（2）主程序 MainActivity.java。

```
1  package （略）
2  import （略）
3  public class MainActivity extends AppCompatActivity
4  {
5    ImageView img;
6    Button connBtn;
7    TextView txt;
8    @Override
9    protected void onCreate(Bundle savedInstanceState) {
10    super.onCreate(savedInstanceState);
11    setContentView(R.layout.activity_main);
12    img = (ImageView)findViewById(R.id.imageView);
13    img.setImageResource(R.drawable.a6);
14    txt = (TextView)findViewById(R.id.textView);
15    connBtn = (Button)findViewById(R.id.button);
16    connBtn.setOnClickListener(new mClick());
17    }
18    class mClick implements View.OnClickListener
19    {
20     String str;
21     @Override
22     public void onClick(View v)
23     {
```

```
24      if(v == connBtn) {
25       connSocket conn = new connSocket();        ── 需要利用 Future 对象获得执行结果
26       try {
27        FutureTask<String> msg = new FutureTask<String>(conn);
28        new Thread(msg).start();
29        str = msg.get();        ── 使用 get()方法获取线程的返回值
30        txt.append(str);
31        } catch (Exception e) { txt.setText("连接错误!!!!!! ");}
32       finally { conn.disConnet(); }
33      } //if_end              //使用get()方法获取线程的返回值
34     } //onClick_end
35    } //mClick_end
36  }
```

(3) 修改 AndroidManifest.xml 的配置（一定不要忘了）。

`<uses-permission android:name="android.permission.INTERNET"/>`

(4) 服务器端程序 Server.java。

```
1   import java.io.DataInputStream;
2   import java.io.DataOutputStream;
3   import java.io.IOException;
4   import java.net.ServerSocket;
5   import java.net.Socket;

6   public class Server {
7    private ServerSocket ss;
8    private Socket socket;
9    private DataInputStream in;
10   private DataOutputStream out;
11   public Server(){
12     new ServerThread().start();
13   }
14   class ServerThread extends Thread{
15    public void run() {
16    try {
17     ss=new ServerSocket(8888);
18     System.out.println("服务器启动了");
19     while(true){
20      socket = ss.accept();
21      System.out.println("有客户端连接到服务器");
22      in = new DataInputStream(socket.getInputStream());
23      out = new DataOutputStream(socket.getOutputStream());
24      String msg = "";
25      if((msg = in.readUTF()) != null){
26         System.out.println(msg);
27      }
28      out.writeUTF("恭喜你，连接服务器成功!   \n");
29      sleep(500);
```

```
30         if((msg=in.readUTF())!=null){
31             System.out.println(msg);
32         }
33      out.writeUTF("你发来的数据服务器收到了。^_^");
34      out.flush();
35     }
36     } catch (IOException | InterruptedException e) {
37             e.printStackTrace();
38     }finally{
39     try {
40         in.close();
41         out.close();
42         } catch (IOException e) {
43             e.printStackTrace();
44         }
45     }
46    }
47   }
48    public static void main(String[] args) throws IOException {
49         new Server();
50     }
51 }
```

6.3 基于 HTTP 协议网络程序设计

6.3.1 建立 PHP 服务器网站

如今有不少人通过手机上网,"哪里有人群,哪里就有发展",这也导致互联网正在向移动端发展。人们使用手机通过无线网络获取 Web 服务器的数据,如图 6.11 所示。

图 6.11 手机获取网络服务器的数据

6.3.2 应用 HttpURLConnection 访问 Web 服务器

1. HttpURLConnection 类

HttpURLConnection 是一种多用途、轻量级的 HTTP 客户端,大多数的应用程序可以使用它来进行 HTTP 操作。但 HttpURLConnection 是 Java 的标准类,没有对其进行封装,需要进行比较复杂的设置,用起来不太方便。

2. StrictMode 类

在 MainActivity 中调用 HttpURLConnection 类的网络操作方法可能会导致 Activity 的

一些问题,在 Android 2.3 版本以后系统增加了 StrictMode 类,这个类对网络的访问方式进行了一定的改变。StrictMode 通常用于捕获磁盘访问或者网络访问中与主进程之间交互产生的问题,因为在主进程中 UI 操作和一些动作的执行是最经常用到的,它们之间会产生一定的冲突问题。将磁盘访问和网络访问从主线程中剥离可以使磁盘或者网络的访问更加流畅,从而提升响应度和用户体验。

在使用 HttpURLConnection 前需要调用 HttpURLConnection 的以下两个方法。

(1) StrictMode.setThreadPolicy():线程对象管理策略。

(2) StrictMode.setVmPolicy():StrictModeVM 虚拟机对象管理策略。

StrictMode 方法需要在主页面的 onCreate 方法中添加以下代码:

```
StrictMode.setThreadPolicy(
  new StrictMode.ThreadPolicy.Builder()     //构造StrictMode线程对象
    .detectDiskReads()                       //当发生磁盘读操作时输出
    .detectDiskWrites()                      //当发生磁盘写操作时输出
    .detectNetwork()                         //访问网络时输出,包括磁盘读/写和网络I/O
    .penaltyLog()                            //以日志的方式输出
    .build()
);
```

setVmPolicy 是关于 VM 虚拟机等方面的策略。

```
StrictMode.setVmPolicy(
  new StrictMode.VmPolicy.Builder()         //构造StrictModeVM虚拟机对象
    .detectLeakedSqlLiteObjects()           //探测SQLite数据库操作
    .detectLeakedClosableObjects()          //探测关闭操作
    .penaltyLog()                            //以日志的方式输出
    .penaltyDeath()
    .build()
);
```

【例 6-7】 从 Web 服务器读取图像文件。

(1) 设计界面布局文件 activity_main.xml。如图 6.12 所示,设置一个按钮两个用于显示信息的文本框、一个显示图像的 ImageView。

图 6.12 从 Web 服务器读取图像文件

(2) 设计主程序 MainActivity.java。

```
1    package com.example.ex6_http;
2    import android.graphics.Bitmap;
3    import android.graphics.BitmapFactory;
4    import android.os.Handler;
5    import android.os.Message;
6    import android.os.StrictMode;
7    import android.support.v7.app.AppCompatActivity;
8    import android.os.Bundle;
9    import android.view.View;
10   import android.widget.Button;
11   import android.widget.EditText;
12   import android.widget.ImageView;
13   import java.io.InputStream;
14   import java.net.HttpURLConnection;
15   import java.net.URL;

16   public class MainActivity extends AppCompatActivity {
17     ImageView img;
18     EditText txt1, txt2;
19     Button connBtn;
20     HttpURLConnection conn = null ;
21     InputStream inStrem = null;
22     String str = "http://58.199.71.24/dukou.jpg";  //使用Web网站IP地址
23     HHandler mHandler = new HHandler();
24     @Override
25     protected void onCreate(Bundle savedInstanceState) {
26       super.onCreate(savedInstanceState);
27       setContentView(R.layout.activity_main);
28       img = (ImageView)findViewById(R.id.imageView);
29       txt1 = (EditText)findViewById(R.id.editText1);
30       txt2 = (EditText)findViewById(R.id.editText2);
31       connBtn = (Button)findViewById(R.id.button);
32       connBtn.setOnClickListener(new mClick());
33     }
34     class mClick implements View.OnClickListener {
35       public void onClick(View arg0) {
36       StrictMode.setThreadPolicy(
37         new StrictMode
38           .ThreadPolicy
39           .Builder()
40          .detectDiskReads()
41          .detectDiskWrites()
42          .detectNetwork()
43          .penaltyLog()
44          .build()
45        );
46     StrictMode.setVmPolicy(
```

```
47      new StrictMode
48        .VmPolicy
49        .Builder()
50        .detectLeakedSqlLiteObjects()
51        .detectLeakedClosableObjects()
52        .penaltyLog()
53        .penaltyDeath()
54        .build()
55      );
56      getPicture();
57    }
58  }
59  private void getPicture(){
60    try {
61      URL url = new URL(str);              //构建图片的URL地址
62      conn = (HttpURLConnection) url.openConnection();
63      conn.setConnectTimeout(5000);        //设置超时的时间, 5000毫秒即5秒
64      conn.setRequestMethod("GET");        //设置获取图片的方式为GET
65      if ( conn.getResponseCode() == 200) { //响应码为200则访问成功
66        //获取连接的输入流,这个输入流就是图片的输入流
67        inStrem = conn.getInputStream();
68        Bitmap bmp= BitmapFactory.decodeStream(inStrem);
69        //由于不是msg,因此不能使用sendMessage(msg)方法
70        mHandler.obtainMessage(0, bmp).sendToTarget();    ← 向Handler发送消息,更新UI
71        int result = inStrem.read();
72        while (result != -1){
73          txt1.setText((char)result);
74          result = inStrem.read();
75        }
76        //关闭输入流
77        inStrem.close();
78        txt1.setText("(1)建立输入流成功! ");
79      }
80    }catch(Exception e2)   { txt1.setText("(3)IO流失败");}
81  }  //getPicture()结束

82  /**
83  Android利用Handler来实现UI线程的更新。
84  Handler是Android中的消息发送器,主要接受子线程发送的数据,并用此数据配合主线程
    更新UI。
85  接受消息,处理消息,此Handler会与当前主线程一块运行
86  */
87  class HHandler extends Handler
88  {  //子类必须重写此方法,接收数据
89    public void handleMessage(Message msg){
90      super.handleMessage( msg);
91      txt2.setText("(2)下载图像成功!");
92      img.setImageBitmap((Bitmap) msg.obj);    //更新UI
93    }
```

```
94      }
95    }    //主类结束
```

（3）修改配置文件，设置网络访问权限。

```
<uses-permission android:name="android.permission.INTERNET" />
```

【例 6-8】 以 GET 方式和 POST 方式向 Web 服务器读取及发送数据。

依题意，在手机客户端编写界面布局文件和主程序文件，在 Web 服务器端编写接收 GET 请求的 play-get.php 文件和接收 POST 请求的 play-post.php 文件。

（1）设计界面布局文件 activity_main.xml。界面布局如图 6.13 所示。

图 6.13 以 GET 方式和 POST 方式向 Web 服务器发送请求

程序代码如下：

```
1    <?xml version="1.0" encoding="utf-8"?>
2    <LinearLayout xmlns:android="http://schemas.android.com/apk/res/android"
3      xmlns:tools="http://schemas.android.com/tools"
4      android:layout_width="match_parent"
5      android:layout_height="match_parent"
6      android:orientation="vertical"
7      tools:context="com.example.ex6_8.MainActivity"
8      android:weightSum="1">
9      <TextView
10        android:layout_width="wrap_content"
11        android:layout_height="wrap_content"
```

```
12      android:text="Http服务"
13      android:id="@+id/textView"
14      android:layout_gravity="center_horizontal"
15      android:textSize="28sp" />
16    <LinearLayout
17      android:orientation="horizontal"
18      android:layout_width="match_parent"
19      android:layout_height="wrap_content">
20      <TextView
21        android:layout_width="wrap_content"
22        android:layout_height="wrap_content"
23        android:text="编号："
24        android:id="@+id/textView2"
25        android:textSize="18sp" />
26      <EditText
27        android:layout_width="wrap_content"
28        android:layout_height="wrap_content"
29        android:id="@+id/editText_ID"
30        android:layout_weight="1" />
31    </LinearLayout>
32    <LinearLayout
33      android:orientation="horizontal"
34      android:layout_width="match_parent"
35      android:layout_height="wrap_content">
36      <TextView
37        android:layout_width="wrap_content"
38        android:layout_height="wrap_content"
39        android:text="姓名："
40        android:id="@+id/textView3"
41        android:textSize="18sp" />
42      <EditText
43        android:layout_width="match_parent"
44        android:layout_height="wrap_content"
45        android:id="@+id/editText_Name" />
46    </LinearLayout>
47    <LinearLayout
48      android:orientation="horizontal"
49      android:layout_width="match_parent"
50      android:layout_height="wrap_content">
51      <TextView
52        android:layout_width="wrap_content"
53        android:layout_height="wrap_content"
54        android:text="邮箱："
55        android:id="@+id/textView4"
56        android:textSize="18sp" />
```

```xml
57      <EditText
58          android:layout_width="match_parent"
59          android:layout_height="wrap_content"
60          android:id="@+id/editText_email" />
61    </LinearLayout>
62    <LinearLayout
63        android:layout_width="match_parent"
64        android:layout_height="wrap_content"
65        android:layout_alignParentTop="true"
66        android:layout_centerHorizontal="true"
67        android:gravity="center_horizontal">
68      <Button
69          android:layout_width="wrap_content"
70          android:layout_height="wrap_content"
71          android:text="Get请求"
72          android:id="@+id/button_get"
73          android:textSize="18sp" />
74      <Button
75          android:layout_width="wrap_content"
76          android:layout_height="wrap_content"
77          android:text="Post请求"
78          android:id="@+id/button_psot"
79          android:textSize="18sp" />
80    </LinearLayout>
81    <TextView
82        android:layout_width="match_parent"
83        android:layout_height="wrap_content"
84        android:text="显示服务器响应信息"
85        android:id="@+id/textView_txt"
86        android:layout_weight="0.05" />
87  </LinearLayout>
```

（2）设计主程序 MainActivity.java。

```java
1   package com.example.ex6_8;
2   import android.os.StrictMode;
3   import android.support.v7.app.AppCompatActivity;
4   import android.os.Bundle;
5   import android.view.View;
6   import android.widget.Button;
7   import android.widget.EditText;
8   import android.widget.TextView;
9   import java.io.BufferedReader;
10  import java.io.InputStreamReader;
11  import java.io.PrintWriter;
12  import java.net.HttpURLConnection;
```

```java
13  import java.net.URL;
14  import java.util.ArrayList;
15  import java.util.HashMap;
16  import java.util.List;
17  import java.util.Map;
18  public class MainActivity extends AppCompatActivity {
19    Button getBtn, postBtn;
20    TextView txt;
21    EditText editsid, editname, editemail;
22    @Override
23    protected void onCreate(Bundle savedInstanceState) {
24      super.onCreate(savedInstanceState);
25      setContentView(R.layout.activity_main);
26      getBtn = (Button)findViewById(R.id.button_get);
27      postBtn = (Button)findViewById(R.id.button_psot);
28      editsid = (EditText)findViewById(R.id.editText_ID);
29      editname = (EditText)findViewById(R.id.editText_Name);
30      editemail = (EditText)findViewById(R.id.editText_email);
31      txt = (TextView)findViewById(R.id.textView_txt);
32      setVersion();    //---设置线程策略---
33      getBtn.setOnClickListener(new mClick());
34      postBtn.setOnClickListener(new mClick());
35    }
36    class mClick implements View.OnClickListener{
37      StringBuilder stringBuilder = new StringBuilder();
38      BufferedReader buffer = null;
39      HttpURLConnection connGET = null;
40      HttpURLConnection connPOST = null;
41      @Override
42      public void onClick(View v) {
43        if(v == getBtn) {
44          //获取界面文本框中的文字内容
45          String sid=editsid.getText().toString();
46          String name=editname.getText().toString();
47          String email=editemail.getText().toString();
48          try{
49            String str = "http://58.199.71.24/test/play-get.php?sid=" +
50                  sid + "&name=" + name + "&email=" + email;
51            URL url = new URL(str);              //构建Web服务器的URL地址
52            connGET = (HttpURLConnection) url.openConnection();
53            connGET.setConnectTimeout(5000);     //设置超时的时间，5000毫秒即5秒
54            connGET.setRequestMethod("GET");     //设置获取数据的方式为GET
55            if ( connGET.getResponseCode() == 200) {
56              buffer = new BufferedReader(new
```

```
57            InputStreamReader(connGET.getInputStream()));
58        for(String s = buffer.readLine();
59                s != null;
60                s = buffer.readLine()){
61            stringBuilder.append(s);          //构造字符串
62        }
63        txt.setText(stringBuilder);
64        buffer.close();
65    }
66  }
67  catch(Exception e){
68      txt.setText("get 提交 err.....");
69  }
70  } //if(v == getBtn)结束
71  if(v == postBtn) {
72    //获取界面文本框中的文字内容
73    String sid=editsid.getText().toString();
74    String name=editname.getText().toString();
75    String email=editemail.getText().toString();
76    try{
77        String str="http://58.199.71.24/test/play-post.php";
78        URL url = new URL(str);           //构建Web服务器的URL地址
79        connPOST = (HttpURLConnection) url.openConnection();
80        connPOST.setConnectTimeout(5000);  //设置超时的时间,5000毫秒即5秒
81        connPOST.setRequestMethod("POST");//设置获取数据的方式为POST
82        //发送POST请求必须设置以下两行
83        connPOST.setDoOutput(true);
84        connPOST.setDoInput(true);
85        //----------发送数据--------
86        //建立对应的输出流
87    PrintWriter printWriter = new PrintWriter(connPOST.getOutputStream());
88    Map<String, Object> paramsMap = new HashMap<String, Object>();
89    paramsMap.put("sid", sid);
90    paramsMap.put("name", name);
91    paramsMap.put("email", email);
92    printWriter.write(paramsMap.toString()); //发送请求参数
93    printWriter.flush(); //flush输出流的缓冲
94        //----------接收数据--------
95        //定义BufferedReader输入流来读取URL的返回数据
96    buffer = new BufferedReader(new
97            InputStreamReader(connPOST.getInputStream()));
98    for(String s = buffer.readLine();
99            s != null;
100           s = buffer.readLine()){
101       stringBuilder.append(s);   //构造字符串
```

```
102        }
103        txt.setText(stringBuilder);
104        buffer.close();
105     } //try结束
106     catch(Exception e){ txt.setText("response err....."); }
107     }
108 }
109 }
110 //设置线程策略
111 void setVersion()
112 {
113   StrictMode.setThreadPolicy(new StrictMode
114     .ThreadPolicy.Builder()
115     .detectDiskReads()
116     .detectDiskWrites()
117     .detectNetwork()  //这里若替换为detectAll()就包括了磁盘读/写和网络I/O
118     .penaltyLog()     //打印logcat,也可以定位到dropbox,通过文件保存log
119     .build());
120   StrictMode.setVmPolicy(new StrictMode
121     .VmPolicy.Builder()
122     .detectLeakedSqlLiteObjects()  //探测SQLite数据库操作
123     .penaltyLog() //打印logcat
124     .penaltyDeath()
125     .build());
126   }
127 }
```

（3）修改配置文件，设置网络访问权限。

`<uses-permission android:name="android.permission.INTERNET" />`

（4）Web 服务器端接收 GET 请求的 play-get.php 文件。

```
1  <?php
2    header("Content-Type: text/html;charset=utf-8");//设置页面显示的文字编码
3    echo "GET: ";
4    print_r($_GET);
5    $ssid=$_GET["sid"];
6    $sname=$_GET["name"];
7    $semail=$_GET["email"];
8    print_r("GET给服务器的值为数组\n" );
9    print_r($ssid );
10   print_r("\n");
11   print_r($sname);
12   print_r("\n");
13   print_r($semail);
```

```
14    print_r("\n");
15  ?>
```

(5) Web 服务器端接收 POST 请求的 play-post.php 文件。

```
1   <?php
2     header("Content-Type: text/html;charset=utf-8");//设置页面显示的文字编码
3     echo "POST:  ";
4     print_r($_POST);
5     $ssid=$_POST["sid"];
6     $sname=$_POST["name"];
7     $semail=$_POST["email"];
8     print_r("POST给服务器的值为数组\n" );
9     print_r($ssid );
10    print_r("\n");
11    print_r($sname);
12    print_r("\n");
13    print_r($semail);
14    print_r("\n");
15  ?>
```

习 题 6

1. 编写一个用户注册程序向远程服务器注册。
2. 编写一个具有密码验证功能的远程用户登录程序。

第 7 章　应用 Volley 框架访问 Web 服务器

7.1　Volley 框架及其应用

7.1.1　Volley 包的下载与安装

1. Volley 简介

在开发 Android 应用项目的时候经常需要用来自网络的数据，通常应用程序都会使用 HTTP 协议来发送和接收网络数据。在 Android 系统中主要用 HttpURLConnection 对象进行 HTTP 通信，我们几乎在任何项目的代码中都能看到这个类的身影，其使用率非常高（早期版本还使用 HttpClient 类，现在已经废弃）。

不过 HttpURLConnection 的用法稍微有些复杂，如果不进行适当封装，很容易就会写出不少重复代码。于是，一些 Android 网络通信框架也就应运而生。

Android 开发团队也意识到有必要将 HTTP 的通信操作再进行简单化，于是在 2013 年推出了一个新的网络通信框架——Volley。Volley 既可以非常简单地进行 HTTP 通信，也可以轻松加载网络上的图片。除了简单、易用之外，Volley 在性能方面也进行了大幅度调整，它的设计目标就是非常适合进行数据量不大但通信频繁的网络操作。

2. 下载和安装 Volley

用户可以到国内网站 "http://download.csdn.net/detail/sinyu890807/7152015" 下载 volley.jar。

新建一个 Android 项目，将 volley.jar 文件复制到 libs 目录下，如图 7.1 所示。

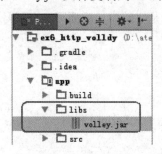

图 7.1　复制 volley 包

右击新粘贴的 volley.jar 项，在弹出的菜单中选择 Add As Library 命令完成 jar 包的安装，如图 7.2 所示。

图 7.2　安装 volley.jar 包

7.1.2　JSON 数据格式简介

Volley 在处理数据的接收与发送时使用的是 JSON 数据格式，下面对 JSON 数据格式进行简要介绍。

JSON（JavaScript Object Notation）是一种轻量级的数据交换格式。JSON 采用完全独立于语言的纯文本格式，易于人们阅读和编写，同时也易于机器解析和生成（一般用于提升网络传输速率），因此 JSON 成为网络传输中理想的数据交换语言。

1. JSON 数据格式

JSON 用"键-值"对形式表示数据，其数据的书写格式如下：

```
键名(key) : 值(value)
```

"键-值"对的键名 key 必须是字符串，后面写一个冒号":"，然后是值 value，值 value 可以是字符串、数值、布尔值。

例如：

```
"firstName" : "John"
```

这很容易理解，等价于下列 JavaScript 语句：

```
firstName = "John"
```

2. JSON 对象

JSON 对象可以包含多个"键-值"对，要求在大括号"{ }"中书写，"键-值"对之间用逗号","分隔。

例如：

```
{ "firstName":"John" , "lastName":"Doe" , "age":20 }
```

这一点也很容易理解，等价于下列 JavaScript 语句：

```
firstName = "John"
lastName = "Doe"
age = 20
```

JSON 对象的值也可以是另一个对象，例如：

```
{
    "Name":"John" ,
    "age": 20 ,
    "hobby":"打篮球",
    "friend":{"Name":"Suny" , "age":19 , "hobby":"看书"}
}
```

3. JSON 数组

JSON 数组可以包含多个 JSON 数据做数组元素，每个元素之间用逗号","分隔，要求在方括号"[ ]"中书写。

例如：

```
var meber = [ "John" , 20 , "打篮球" ];
```

JSON 数组的元素可以包含多个对象，例如：

```
var employees = [
    { "firstName":"John" , "lastName":"Doe" },
    { "firstName":"Anna" , "lastName":"Smith" },
    { "firstName":"Peter" , "lastName":"Jones" }
];
```

可以像这样访问 JavaScript 对象数组中的第一个元素：

```
employees[0].lastName;
```

其返回值为 Doe。

用户也可以修改其数据：

```
employees[0].lastName = "Jobs";
```

4. JSON 文件

JSON 文件的类型是".json"，可以用记事本或其他编辑工具编写 JSON 文件。

5. 解析 JSON 数据

Android 解析 JSON 格式数据需要使用 JSONObject 对象和 JSONArray 对象，下面通过一个示例说明 Android 解析 JSON 格式数据的方法。

【例 7-1】 解析 JSON 格式数据示例。

(1) 界面布局如图 7.3 所示。

图 7.3　解析 JSON 格式数据的界面布局

(2) 控制文件 MainActivity.java 的代码如下：

```
1  package com.example.ex7_json;
2  import android.support.v7.app.AppCompatActivity;
3  import android.os.Bundle;
4  import android.view.View;
5  import android.widget.Button;
6  import android.widget.EditText;
7  import org.json.JSONArray;
8  import org.json.JSONException;
9  import org.json.JSONObject;
10
11 public class MainActivity extends AppCompatActivity {
12    EditText txt1, txt2, txt3;
13    Button   jsonBtn, arrayBtn;
14    /* 设有JSON数据
15     JSONArray jdata = [{"sid":1001, "name":"张大山"},
16                {"sid":1002, "name":"李小丽"} ];
17    */
18    JSONObject jid,jname;
19    @Override
20    protected void onCreate(Bundle savedInstanceState) {
21       super.onCreate(savedInstanceState);
```

```
22      setContentView(R.layout.activity_main);
23      txt1 = (EditText)findViewById(R.id.editText);
24      txt2 = (EditText)findViewById(R.id.editText2);
25      txt3 = (EditText)findViewById(R.id.editText3);
26.     jsonBtn = (Button)findViewById(R.id.button);
27      arrayBtn = (Button)findViewById(R.id.button2);
28      jsonBtn.setOnClickListener(new mClick());
29      arrayBtn.setOnClickListener(new mClick());
30     }
31
32    void setJsonData() {
33       try {
34           jid = new JSONObject();
35           jname = new JSONObject();
36           jid.put("sid",1001);
37           jname.put("name","张大山");
38           String sid=jid.getString("sid");               设置单条数据
39           txt1.setText(sid);
40           String sname = jname.getString("name");
41           txt2.setText(sname);
42          }catch (JSONException e){  }
43    }
44
45    void setArrayData(){
46       try {
47           JSONArray jdata = new JSONArray();
48           JSONObject p1 = new JSONObject();
49           JSONObject p2 = new JSONObject();
50           p1.put("sid",1001).put("name","张大山");        设置JSON数组元素
51           p2.put("sid",1002).put("name","李小丽");
52           jdata.put(p1);
53           jdata.put(p2);
54          String sid , sname;
55           int length = jdata.length();
56           for(int i=0; i<length; i++){      //遍历JSONArray
57            JSONObject jsonObject = jdata.getJSONObject(i);
58             sid = jsonObject.getString("sid") + ":";
59             sname = jsonObject.getString("name") + "\n";
60             txt3.append(sid + sname);
61           }
62       }catch (JSONException e){  }
63    }
64
65   public class mClick implements View.OnClickListener
66    {
```

```
67        @Override
68        public void onClick(View v) {
69            if(v == jsonBtn)   setJsonData();
70            else if(v == arrayBtn) setArrayData();
71        }
72    }
73 }
```

程序的运行结果如图 7.4 所示。

图 7.4 解析 JSON 数据程序的运行结果

7.1.3 Volley 的工作原理和几个重要对象

1. Volley 的基本工作原理

Volley 在工作时首先由主线程（应用程序）发起一条 HTTP 请求，将请求添加到缓存队列中，然后通过缓存调度线程从缓存队列中取出一个请求，在缓存中解析并做出响应，最后将解析后的响应发送给主线程。在 Volley 内部创建两个线程，一个为缓存调度线程，另一个为网络调度线程，优先在缓存中解析并响应请求，如果缓存不能解析，则由网络线程使用 HTTP 发送请求给远程 Web 服务器解析。Volley 的基本工作原理如图 7.5 所示。

2. Volley 的几个重要对象

使用 Volley 框架需要创建几个重要对象，现介绍如下。
（1）RequestQueue：用来执行请求的请求队列。
（2）Request：用来构造一个请求对象。

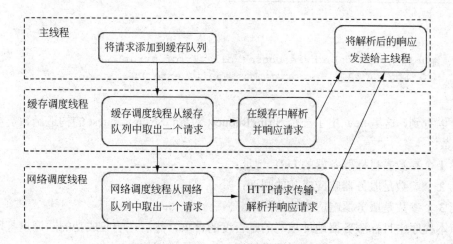

图 7.5 Volley 框架的基本工作原理

Request 对象主要有以下几种类型。
- StringRequest：响应的主体为字符串。
- JsonArrayRequest：发送和接收 JSON 数组。
- JsonObjectRequest：发送和接收 JSON 对象。
- ImageRequest：发送和接收 Image 图像对象。

7.1.4 Volley 的基本使用方法

Volley 的用法非常简单，假设应用程序发起一条 HTTP 请求，然后接收 HTTP 响应，其步骤如下。

（1）创建一个 RequestQueue 对象，可以调用以下方法得到：

```
RequestQueue mQueue = Volley.newRequestQueue(context);
```

注意这里得到的 RequestQueue 是一个请求队列对象，它可以缓存所有的 HTTP 请求，然后按照一定的算法并发地发出这些请求。

RequestQueue 内部的设计是非常适合高并发的，因此不必为每一次 HTTP 请求都创建一个 RequestQueue 对象，否则非常浪费资源，基本上在每一个需要和网络交互的 Activity 中创建一个 RequestQueue 对象就可以了。

（2）为了要发出一条 HTTP 请求，还需要创建一个 StringRequest 对象，例如：

```
StringRequest stringRequest = new StringRequest(
    "http://www.baidu.com",              //第1个参数
    new Response.Listener<String>()      //第2个参数
    {
     @Override
     public void onResponse(String response)
                                         //onResponse获取到服务器响应的值
        {  Log.d("TAG", response);  }
    },
new Response.ErrorListener()    //第3个参数
```

```
{
    @Override
    public void onErrorResponse(VolleyError error)
      { Log.e("TAG", error.getMessage(), error);    }
});
```

可以看到，这里 new 出了一个 StringRequest 对象，StringRequest 的构造函数需要传入3个参数：

第1个参数是目标服务器的 URL 地址；

第2个参数是服务器响应成功的回调；

第3个参数是服务器响应失败的回调。

在本程序中目标服务器地址填写的是百度的首页，然后在响应成功的回调里打印出服务器返回的内容，在响应失败的回调里打印出失败的详细信息。

（3）将这个 StringRequest 对象添加到 RequestQueue 里面，例如：

mQueue.add(stringRequest);

注意：使用 Volley 框架一定要在 AndroidManifest.xml 文件中添加网络权限：

```
<uses-permission android:name="android.permission.INTERNET" />
```

【**例 7-2**】 应用 Volley 框架从 Web 服务器的 JSON 文件中读取数据。

该程序分为手机端应用程序和服务器端的 JSON 文件，现设计如下：

（1）在 Web 服务器端建立 jsonData.json 文件。在 Web 服务器的 www 根目录下新建 test 目录，并在 test 目录中建立 jsonData.json 文件，其数据内容如下：

[{"sid":1001, "name":"张大山"},{"sid":1002, "name":"李小丽"}]

（2）把 volley.jar 复制到项目的 app\libs 目录下，并完成 jar 包的安装。

（3）手机端界面布局程序设计。手机端界面布局程序如图 7.6 所示。

图 7.6　界面布局设计

(4)主程序 MainActivity.java 设计。

```java
1  package com.example.ex7_volldy;
2  import android.support.v7.app.AppCompatActivity;
3  import android.os.Bundle;
4  import android.util.Log;
5  import android.view.View;
6  import android.widget.Button;
7  import android.widget.EditText;
8  import com.android.volley.RequestQueue;
9  import com.android.volley.Response;
10 import com.android.volley.VolleyError;
11 import com.android.volley.toolbox.StringRequest;
12 import com.android.volley.toolbox.Volley;
13 public class MainActivity extends Activity {
14   Button   volleyBtn;
15   EditText txt;
16   @Override
17   protected void onCreate(Bundle savedInstanceState) {
18     super.onCreate(savedInstanceState);
19     setContentView(R.layout.activity_main);
20     volleyBtn=(Button)findViewById(R.id.button2);
21     txt = (EditText)findViewById(R.id.editText);
22     volleyBtn.setOnClickListener(new mClick());
23   }
24   class mClick implements View.OnClickListener
25   {
26     String str;
27     @Override
28     public void onClick(View v) {
29     if(v == volleyBtn){
30       String URL = "http://192.168.1.1/test/jsonData.json";
31       RequestQueue mQueue = Volley.newRequestQueue(MainActivity.this);
32       StringRequest stringRequest = new  StringRequest(
33           URL,              //第1个参数,Web服务器的IP地址
34         new Response.Listener<String>() {  //第2个参数,Volley的监听器
35           @Override
36           //onResponse()方法获取接收到的数据值
37           public void onResponse(String response)
38           { txt.setText(response); }
39         },
40         new Response.ErrorListener() {  //第3个参数,响应失败
41           @Override
42           public void onErrorResponse(VolleyError error)
43           { Log.e("TAG", error.getMessage(), error); }
```

```
44          })
45  {      //将Volley默认的ISO-8859-1格式转换为utf-8格式
46      @Override
47      protected Response<String> parseNetworkResponse(
48       NetworkResponse response){
49       try {   //jsonObject要和前面的类型一致,此处都是String类型
50           String jsonString = new String(response.data, "utf-8");
51           return Response.success(jsonString,
52                   HttpHeaderParser.parseCacheHeaders(response));
53       } catch (UnsupportedEncodingException e) {
54           return Response.error(new ParseError(e));
55       } catch (Exception je) {
56           return Response.error(new ParseError(je));
57       }
58      }
59  } ;
60          mQueue.add(stringRequest);
61        } //if_end
62      } //onClick()_end
63    } //class mClick_end
64  } //class MainActivity_end
```

（第45~59行为"解决汉字乱码"）

由于 Volley 默认编码为 ISO-8859-1，因此显示汉字会出现乱码，程序的第 45~59 行进行转码，将 ISO-8859-1 的编码格式转换为 utf-8 格式，从而能正确地显示汉字。

（5）修改配置文件，添加访问网络权限：

```
<uses-permission android:name="android.permission.INTERNET" />
```

运行程序时首先启动 Web 服务器，然后运行手机端应用程序。程序运行结果如图 7.7 所示。

图 7.7 应用 Volley 远程 Web 服务器读取 JSON 数据

7.2 应用 Volley 框架设计网络音乐播放器

下面介绍如何应用 Volley 框架设计网络音乐播放器。

【例 7-3】 音乐信息保存在远程服务器的一个 JOSN 格式文件中，读取 JSON 数据内容，把音乐的相关信息显示到屏幕上，并播放音乐。

（1）将 volley.jar 包复制到项目的 libs 目录之下，右击新粘贴的 volley.jar 项，在弹出的菜单中选择 Add As Library 命令安装 volley.jar 包。

（2）将一个音乐文件 mtest.mp3 复制到 Web 服务器根目录 www\music 下。现用记事本新建一个 JSON 格式数据文件用于存放音乐资源信息，文件保存到远程 Web 服务器（www\music\music_info.json）中，其数据如下：

{"name":"伤不起","singer":"王麟","mp3":"mtest.mp3"}

注意：music_info.json 需要保存为 utf-8 格式。在记事本中选择"另存为"命令，将编码格式更改为 utf-8 即可。

（3）读取远程服务器文件内容的核心语句。

```
RequestQueue mQueue = Volley.newRequestQueue(MainActivity.this);
JsonObjectRequest jsonObjectRequest = new JsonObjectRequest(
    jsonUrl,      //第1个参数，请求的网址
    null,         //第2个参数
    new Response.Listener<JSONObject>() {  //第3个参数，响应正确时的处理
        @Override
        public void onResponse(JSONObject response) {
        try {
            String sname=new String(response.getString("name"));
            String ssinger=new String(response.getString("singer"));
            String smp3=new String(response.getString("mp3"));
            musicinfo.setName(sname);
            musicinfo.setSinger(ssinger);
            musicinfo.setMp3(smp3);
        }catch (JSONException e){ Log.e("json错误", e.getMessage(), e);
        }
        }
    },
    new Response.ErrorListener() {    //第4个参数，错误时反馈信息
        @Override
        public void onErrorResponse(VolleyError error)
        { Log.e("TAG错误", error.getMessage(), error); }
    }
);
mQueue.add(jsonObjectRequest);
```

（4）界面布局设计。

在界面布局文件中设置了一个显示音乐信息的列表组件 listView，另外两个按钮分别用于读取远程服务器上的音乐信息文件及播放音乐，并设置 linearLayout(horizontal)的"gravity"属性值为"center"，使这两个按钮居中，如图 7.8 所示。

图 7.8　远程服务器音乐播放器界面布局

（5）修改配置文件，设置网络访问权限。

```
<uses-permission android:name="android.permission.INTERNET" />
```

（6）编写存放和获取网络数据的 JavaBen 文件 Musicinfo.java。

```
1   package com.example.ex7_Music;
2   public class Musicinfo
3   {
4     private String name;
5     private String singer;
6     private String mp3;
7     public void setName(String name){this.name = name;}
8     public void setSinger(String singer){this.singer = singer;}
9     public void setMp3(String mp3){this.mp3 = mp3;}
10    public String getName(){return name;}
11    public String getSinger(){return singer;}
12    public String getMp3(){return mp3;}
```

13 }

(7) 编写主程序 MainActivity.java。

```java
1   package com.example.ex7_Music;
2   import android.media.MediaPlayer;
3   import android.support.v7.app.AppCompatActivity;
4   import android.os.Bundle;
5   import android.util.Log;
6   import android.view.View;
7   import android.widget.ArrayAdapter;
8   import android.widget.Button;
9   import android.widget.ListView;
10  import android.widget.TextView;
11  import com.android.volley.RequestQueue;
12  import com.android.volley.Response;
13  import com.android.volley.VolleyError;
14  import com.android.volley.toolbox.JsonObjectRequest;
15  import com.android.volley.toolbox.Volley;
16  import org.json.JSONException;
17  import org.json.JSONObject;
18  import java.io.IOException;
19  import java.io.UnsupportedEncodingException;

20  public class MainActivity extends    AppCompatActivity {
21      TextView txt;
22      ListView  musicView;    //定义列表组件,用于显示音乐信息
23      ArrayAdapter<String>  adapter;    //定义适配器
24      Button  readBtn, playBtn, stopBtn;
25      String[] list={"","",""};
26      Musicinfo musicinfo = new Musicinfo();   //创建输入/输出数据对象
27      String musicUil="http://58.199.89.161/music/";
28      @Override
29      protected void onCreate(Bundle savedInstanceState) {
30          super.onCreate(savedInstanceState);
31          setContentView(R.layout.activity_main);
32          txt = (TextView)findViewById(R.id.textView2);
33          musicView = (ListView)findViewById(R.id.listView);
34          readBtn = (Button)findViewById(R.id.button);
35          playBtn = (Button)findViewById(R.id.button2);
36          stopBtn = (Button)findViewById(R.id.button3);
37          getServerData();
38          readBtn.setOnClickListener(new mread());
39          playBtn.setOnClickListener(new mplay());
```

```java
40      stopBtn.setOnClickListener(new mstop());
41    }

42    class mread implements View.OnClickListener {
43      public void onClick(View v) {
44       list[0] = "音乐名称: " + musicinfo.getName();
45       list[1] = "歌手: " + musicinfo.getSinger();
46       list[2] = "音乐文件名: " + musicinfo.getMp3();
47       adapter = new ArrayAdapter<String>(
48         MainActivity.this,
49         android.R.layout.simple_list_item_1,
50         list);
51       musicView.setAdapter(adapter);
52      }
53    }
54    class mplay implements View.OnClickListener{
55      public void onClick(View v) {
56       try {
57         String path = musicUil + musicinfo.getMp3();
58         txt.setText(path);
59         MediaPlayer mMediaPlayer = new MediaPlayer();
60         mMediaPlayer.reset();
61         mMediaPlayer.setDataSource(path);
62         mMediaPlayer.prepare();
63         mMediaPlayer.start();
64       }catch (IOException e){    }
65      }
66    }
67    class mstop implements View.OnClickListener{
68     @Override
69      public void onClick(View v) {
70       if (mMediaPlayer.isPlaying()){
71         mMediaPlayer.stop();
72         mMediaPlayer.reset();
73         mMediaPlayer.release();
74       }
75     }
76    }

77    public void getServerData(){
78     String jsonUrl = musicUil + "music_info.json";
79     RequestQueue mQueue = Volley.newRequestQueue(MainActivity.this);
80     JsonObjectRequest jsonObjectRequest = new JsonObjectRequest(
81       jsonUrl,    //第1个参数, 请求的网址
```

```
82      null,           //第2个参数
83      new Response.Listener<JSONObject>() { //第3个参数,响应正确时的处理
84        @Override
85        public void onResponse(JSONObject response) {
86         try {
87           String sname=new String(response.getString("name"));
88           String ssinger=new String(response.getString("singer"));
89           String smp3=new String(response.getString("mp3"));
90           musicinfo.setName(sname);
91           musicinfo.setSinger(ssinger);
92           musicinfo.setMp3(smp3);
93         }catch (JSONException e){ txt.setText("list错误");
94         Log.e("json错误", e.getMessage(), e);
95         }
96       }
97      },
98      new Response.ErrorListener() {      //第4个参数,错误时反馈信息
99       @Override
100       public void onErrorResponse(VolleyError error)
101       { Log.e("TAG错误", error.getMessage(), error); }
102      } )
103     { //将Volley默认的ISO-8859-1格式转换为utf-8格式
104       @Override
105       protected Response<JSONObject> parseNetworkResponse(
106        NetworkResponse response) {
107        try {  //jsonObject要和前面的类型一致,此处都是JSONObject
108          JSONObject jsonObject = new JSONObject(
109            new String(response.data, "utf-8"));
110          return Response.success(jsonObject,
111           HttpHeaderParser.parseCacheHeaders(response));
112        } catch (UnsupportedEncodingException e) {
113          return Response.error(new ParseError(e));
114        } catch (Exception je) {
115         return Response.error(new ParseError(je)) ;
116       }
117      }
118     };
119     mQueue.add(jsonObjectRequest);
120    }
121 }
```

> 解决汉字乱码

程序的运行结果如图7.9所示。

图 7.9 网络音乐播放器

习 题 7

1. 设有 JSON 数组 [{"sid":1001, "name":"张大山"}, {"sid":1002, "name":"李小丽"}]，编写一个通过列表组件 ListView 显示 JSON 数组数据的程序。
2. 设计一个网络音乐播放器，通过点击列表中的歌曲名称播放所选取的音乐。

第 8 章　数据存储技术

本章介绍 Android 系统的数据存储方式。Android 系统提供了多种数据存储方式，有 SQLite 数据库存储方式、文件存储方式、XML 文件的 SharedPreference 存储方式等。

8.1　SQLite 数据库

8.1.1　SQLite 数据库简介

SQLite 数据库是一个关系型数据库，因为它很小，引擎本身只有一个大小不到 300 KB 的文件，所以常作为嵌入式数据库内嵌在应用程序中。SQLite 生成的数据库文件是一个普通的磁盘文件，可以放置在任何目录下。SQLite 是用 C 语言开发的，开放源代码，支持跨平台，最大支持 2048 GB 数据，并且被所有的主流编程语言支持。可以说 SQLite 是一个非常优秀的嵌入式数据库。

SQLite 数据库的管理工具很多，比较常用的有 SQLite Expert Professional，其功能强大，几乎可以在可视化的环境下完成所有数据库操作。用户使用它可以方便地创建数据表和对数据记录进行增加、删除、修改、查询操作。SQLite Expert Professional 的运行界面如图 8.1 所示。

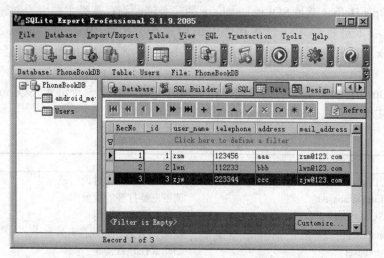

图 8.1　SQLite Expert Professional 的运行界面

在 Android 系统的内部集成了 SQLite 数据库，所以 Android 应用程序可以很方便地使用 SQLite 数据库来存储数据。

用户可以通过 Android 系统的 DDMS 工具将数据库文件复制到本地计算机上，应用 SQLite Expert Professional 对数据库进行操作，完成后再通过 DDMS 工具放回到设备中，当需要操作大量数据时这是比较方便的方法。

8.1.2 管理和操作 SQLite 数据库的对象

Android 提供了创建和使用 SQLite 数据库的 API（Application Programming Interface，应用程序编程接口）。在 Android 系统中主要由 SQLiteDatabase 和 SQLiteOpenHelper 类对 SQLite 数据库进行管理和操作，下面分别对它们进行介绍。

1. SQLiteDatabase 类

在 Android 中主要由 SQLiteDatabase 对象对 SQLite 数据库进行管理，SQLiteDatabase 提供了一系列操作数据库的方法，可以对数据库进行创建、删除、执行 SQL 命令等操作，其常用的方法见表 8-1。

表 8-1 SQLiteDatabase 的常用方法

方法	说明
openOrCreateDatabase(String path, SQLiteDatabase.CursorFactory factory)	打开或创建数据库
openDatabase(String path, SQLiteDatabase.CursorFactory factory，int flags)	打开指定的数据库
insert(String table, String nullColumnHack, ContentValues values)	新增一条记录
delete(String table,String whereClause, String[] whereArgs)	删除一条记录
query(String table,String[] columns, String selection, String[]selectionArgs, String groupBy, String having, String orderBy)	查询一条记录
update(String table,ContentValues values, String whereClause, String[] whereArgs)	修改记录
execSQL(String sql)	执行一条 SQL 语句
close()	关闭数据库

2. SQLiteOpenHelper 类

SQLiteOpenHelper 类是 SQLiteDatabase 的一个辅助类，该类主要用于创建数据库，并对数据库的版本进行管理。当在程序中调用这个类的 getWritableDatabase()方法或者 getReadableDatabase()方法的时候，如果数据库不存在，那么 Android 系统就会自动创建一个数据库。

SQLiteOpenHelper 是一个抽象类，在使用时一般要定义一个继承 SQLiteOpenHelper 的子类，并实现其方法。SQLiteOpenHelper 的常用方法见表 8-2。

表 8-2 SQLiteOpenHelper 的常用方法

方法	说明
onCreate（SQLiteDatabase）	首次生成数据库时调用该方法
onOpen（SQLiteDatabase）	调用已经打开的数据库
onUpgrade（SQLiteDatabase，int，int）	升级数据库时调用
getWritableDatabase()	以读/写方式创建或打开数据库
getReadableDatabase()	创建或打开数据库

8.1.3 SQLite 数据库的操作命令

对数据库的操作有 3 个层次，各层次的操作内容如下。

- 对数据库操作：建立数据库或删除数据库。
- 对数据表操作：建立、修改或删除数据库中的数据表。
- 对记录操作：对数据表中的数据记录进行添加、删除、修改、查询等操作。

下面按上述 3 个层次讲述 SQLite 数据库操作命令的使用方法。

1. 创建及删除数据库

1）创建数据库

创建数据库的方法有多种，可以应用 SQLiteDatabase 对象的 openDatabase()方法及 openOrCreateDatabase()方法创建数据库也可以应用 SQLiteOpenHelper 的子类创建数据库；还可以应用 Activity 继承于父类 android.content.Context 创建数据库的方法 openOrCreateDatabase()来创建数据库。

Context 类的 openOrCreateDatabase（name，mode，factory）方法有 3 个参数：

第 1 个参数 name 为数据库名称。

第 2 个参数 mode 为打开或创建数据库的模式，其模式如下。

- MODE_PRIVATE：只可访问或调用模式，这是默认的模式。
- MODE_WORLD_READABLE：只读模式。
- MODE_WORLD_WRITEABLE：只写模式。

第 3 个参数 factory 为查询数据的游标，通常为 null。

例如要创建一个名称为 PhoneBook.db 的数据库，其数据库的结构如下：

```
String TABLE_NAME = "Users";              //数据表名
String ID = "_id";                        //ID
String USER_NAME = "user_name";           //用户名
String ADDRESS = "address";               //地址
String TELEPHONE = "telephone";           //联系电话
String MAIL_ADDRESS = "mail_address";     //电子邮箱
```

则用 Activity 的 openOrCreateDatabase()方法创建数据库的代码如下：

```
SQLiteDatabase db;
String db_name = "PhoneBook.db";
String sqlStr =
        "CREATE TABLE " + TABLE_NAME + " ("
            + ID + " INTEGER primary key autoincrement, "
            + USER_NAME + " text not null, "
            + TELEPHONE + " text not null, "
            + ADDRESS + " text not null, "
            + MAIL_ADDRESS + " text not null "+ ");";
int mode = Context.MODE_PRIVATE;
db = this.openOrCreateDatabase(Database_name, mode, null);   //创建数据库
db.execSQL(sqlStr);   //执行创建数据库的 SQL 语句
```

（创建数据表的 SQL 语句）

这时，通过 DDMS 可以看到在 data\data\xxxx（包名）\databases 下创建了数据库 PhoneBook.db。

2)删除数据库

当要删除一个指定的数据库文件时需要应用 android.content.Context 类的 deleteDatabase(String name)方法来删除这个指定的数据库。

例如要删除名为 PhoneBook.db 的数据库,则可以使用下列代码:

```
MainActivity.this.deleteDatabase("PhoneBook.db");
```

【例 8-1】 编写一个创建与删除数据库的演示程序。

(1)用户界面设计。在程序的用户界面中设置一个文本标签和两个按钮,如图 8.2 所示。

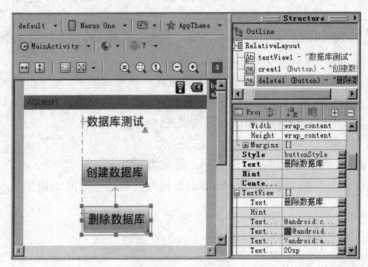

图 8.2 数据库测试程序的用户界面设计

(2)代码设计。

```
1   package com.ex8_1;
2   import android.os.Bundle;
3   import android.app.Activity;
4   import android.content.Context;
5   import android.database.sqlite.SQLiteDatabase;
6   import android.view.View;
7   import android.view.View.OnClickListener;
8   import android.widget.Button;
9   public class MainActivity extends Activity
10  {
11      Button creatBtn, deleteBtn;
12      @Override
13      public void onCreate(Bundle savedInstanceState)
14      {
15          super.onCreate(savedInstanceState);
16          setContentView(R.layout.activity_main);
```

```
17      creatBtn=(Button)findViewById(R.id.creat1);
18      creatBtn.setOnClickListener(new mClick());
19      deleteBtn=(Button)findViewById(R.id.delete1);
20      deleteBtn.setOnClickListener(new mClick());
21   }
22   class mClick implements OnClickListener
23   {
24      @Override
25      public void onClick(View arg0)
26      {
27         if(arg0 == creatBtn)
28         {
29            DBCreate db = new DBCreate();     ← 实例化创建数据库类
30         }
31         else if(arg0 == deleteBtn)
32         {
33         deleteDatabase(DBCreate.Database_name);   ← 删除数据库
34         }
35      }
36   }

37   class DBCreate
38   {
39      static final String Database_name = "PhoneBook.db";  ← 数据库名
40      private DBCreate()
41      {
42         SQLiteDatabase db;
43         String TABLE_NAME = "Users";          //数据表名
44         String ID = "_id";                    //ID
45         String USER_NAME = "user_name";       //用户名      ┐ 定义数据
46         String ADDRESS = "address";           //地址        │ 库结构
47         String TELEPHONE = "telephone";       //联系电话    │
48         String MAIL_ADDRESS = "mail_address"; //电子邮箱    ┘
49         String DATABASE_CREATE =
50            "CREATE TABLE " + TABLE_NAME + " ("
51            + ID + " INTEGER primary key autoincrement,"    ┐ 创建数据
52            + USER_NAME + " text not null, "                │ 表的SQL
53            + TELEPHONE + " text not null, "                │ 语句
54            + ADDRESS + " text not null, "                  │
55            + MAIL_ADDRESS + " text not null "+ ");";       ┘
56         int mode = Context.MODE_PRIVATE;
57         db = openOrCreateDatabase(Database_name, mode, null);  ← 创建数据库
58         db.execSQL(DATABASE_CREATE);
59      }
60   }
```

61 }

运行程序，当单击"创建数据库"按钮后通过 DDMS 工具调试监控视图，在 data\data\
××××（包名）\databases 下可以看到创建的数据库文件 PhoneBook.db，如图 8.3 所示。
之后单击"删除数据库"按钮，数据库文件则被删除。

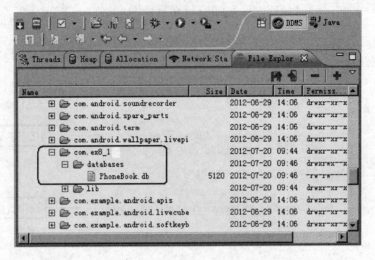

图 8.3　在 DDMS 视图中查看创建的数据库文件 PhoneBook.db

2. 数据表操作

1）创建数据表

创建数据表的步骤如下：

（1）用 SQL 语句编写创建数据表的命令。

（2）调用 SQLiteDatabase 的 execSQL()方法执行 SQL 语句。

例如，在上面创建的数据库中建立一个名为 Users 的数据表。

2）删除数据表

删除数据表的步骤与创建数据表类似，先编写删除表的 SQL 语句，再调用 execSQL()
方法执行 SQL 语句。

例如要删除名为 Users 的数据表，则

```
String sql ="DROP TABLE Users";    ← 删除数据表 Users 的 SQL 语句
db.execSQL(sql);    ← 执行 SQL 语句
```

3. 对数据记录的操作

在数据表中把数据表的列称为字段，把数据表的每一行称为记录。对数据表中的数据
进行操作处理主要是对其记录进行操作处理。

对数据记录的操作处理有两种方法，一种方法是编写一条对记录进行增、删、改、查
的 SQL 语句，通过 exeSQL()方法来执行；另一种方法是使用 Android 系统 SQLiteDatabase
对象的相应方法进行操作。对于使用 SQL 语句执行相应操作的办法，一般数据库的书籍均
有介绍，这里就不再赘述。下面仅介绍 SQLiteDatabase 对象操作数据记录的方法。

1）新增记录

新增记录使用 SQLiteDatabase 对象的 insert()方法。

insert(String table, String nullColumnHack, ContentValues values)方法中的 3 个参数的含义如下。

- 第 1 个参数 table：增加记录的数据表。
- 第 2 个参数 nullColumnHack：空列的默认值，通常为 null。
- 第 3 个参数 ContentValues：为 ContentValues 对象，其实就是一个"键-值"对的字段名称，键名为表中的字段名，键值为要增加的记录数据值。通过 ContentValues 对象的 put()方法把数据存放到 ContentValues 对象中。

例如，下列代码分别把 4 个"键-值"对数据存放到 values 对象中，其键名分别为 USER_NAME（用户名）、TELEPHONE（联系电话）、ADDRESS（住址）、MAIL_ADDRESS（电子邮箱）。

```
ContentValues values = new ContentValues();
values.put(USER_NAME, "张大山");
values.put(TELEPHONE, "13800012345");
values.put(ADDRESS, "南海仙人岛");
values.put(MAIL_ADDRESS, "abc@123.com");
db.insert("Users", null, values);
```

2）修改记录

修改记录使用 SQLiteDatabase 对象的 update()方法。在 update(String table, ContentValues values, String whereClause, String[] whereArgs) 方法中有 4 个参数，其含义如下。

- 第 1 个参数 table：修改记录的数据表。
- 第 2 个参数 ContentValues：ContentValues 对象，存放已做修改的数据的对象。
- 第 3 个参数 whereClause：修改数据的条件，相当于 SQL 语句的 where 子句。
- 第 4 个参数 whereArgs：修改数据值的数组。

例如，下列代码把用户名为张大山的经过修改的数据记录（替换数据表中原来的数据记录）存放到 Values 中，其他数据值不变：

```
ContentValues values = new ContentValues();
values.put(TELEPHONE, "13811011011");
values.put(ADDRESS, "东海彭湖湾");
String where ="USER_NAME = 张大山";
db.update("Users", values, where ,null);
```

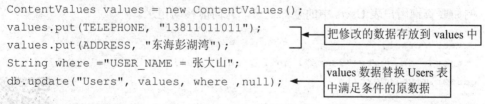

3）删除记录

删除记录使用 SQLiteDatabase 对象的 delete()方法。在 delete(String table, String whereClause, String[] whereArgs)方法中有 3 个参数，其含义如下。

- 第 1 个参数 table：修改记录的数据表。
- 第 2 个参数 whereClause：删除数据的条件，相当于 SQL 语句的 where 子句。
- 第 3 个参数 whereArgs：删除条件的参数数组。

例如，下列代码把用户名为张大山的所有数据删除，即删除条件为"USER_NAME = 张大山"的记录：

```
String where ="USER_NAME = 张大山";
db.delete("Users", where ,null);        ← 删除 Users 表中满足条件的记录
```

4）查询记录

在数据库的操作命令中查询数据的命令是最丰富、最复杂的。在 SQLite 数据库中使用 SQLiteDatabase 对象的 query()方法查询数据。query(String table, String[] columns, String selection, String[] selectionArgs, String groupBy, String having, String orderBy)方法有 7 个参数，其含义如下。

- 第 1 个参数 table：查询记录的数据表。
- 第 2 个参数 columns：查询的字段，如果为 null，则为所有字段。
- 第 3 个参数 selection：查询条件，可以使用通配符"？"。
- 第 4 个参数 selectionArgs：参数数组，用于替换查询条件中的"？"。
- 第 5 个参数 groupBy：查询结果按指定字段分组。
- 第 6 个参数 having：限定分组的条件。
- 第 7 个参数 orderBy：查询结果的排序条件。

5）对查询结果 cursor 的处理

query()方法查询的数据均封装到查询结果 Cursor 对象之中，Cursor 相当于 SQL 语句中 resultSet 结果集上的一个游标，可以向前或向后移动。Cursor 对象的常用方法如下。

- moveToFirst()：移动到第 1 行。
- moveToLast()：移动到最后一行。
- moveToNext()：向前移动一行。
- MoveToPrevious()：向后移动一行。
- moveToPosition(positon)：移动到指定位置。
- isBeforeFirst()：判断是否指向第 1 条记录之前。
- isAfterLast()：判断是否指向最后一条记录之后。

例如要查询用户表 Users 中的全部记录，并向前移动记录。

```
Cursor cursor;
cursor = db.query("Users", null , null, null, null, null, null);
cursor.moveToNext();
```

【例 8-2】 编写程序，建立一个通讯录，可以向前、向后浏览数据记录，也可以添加、修改、删除数据。

（1）用户界面设计。在界面设计中用一个垂直线性布局嵌套了 3 个水平线性布局和 1 个表格布局，从而把整个界面划分为 4 个部分。在第 1 个水平线性布局中嵌套了"建立数据库"和"打开数据库"按钮；在第 2 个水平线性布局中嵌套了"浏览通讯录"文本标签以及"建立数据库"和"打开数据库"按钮；在第 3 个水平线性布局中嵌套

了"添加""修改""删除"和"关闭通讯录"按钮；在表格布局中设置表格为 4 行 2 列，每一行均包含一个文本标签和一个文本编辑框。

用户界面布局设计如图 8.4 所示。

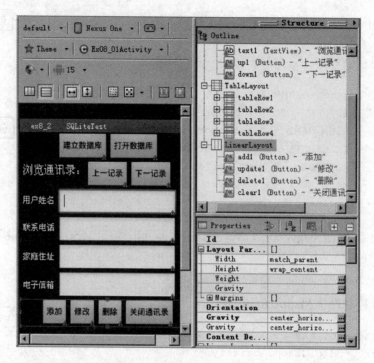

图 8.4 通讯录的界面布局设计

（2）设计数据库程序 DBConnection.java。

```
1   package com.ex8_2;
2   import android.content.Context;
3   import android.database.Cursor;
4   import android.database.sqlite.SQLiteDatabase;
5   import android.database.sqlite.SQLiteOpenHelper;
6
7   public class DBConnection extends SQLiteOpenHelper
8   {
9       static final String Database_name = "PhoneBook.db";    ← 定义数据库名
10      static final int Database_Version = 1;
11      SQLiteDatabase db;    ← 定义 SQLiteDatabase 数据库对象
12      public int id_this;
13      Cursor cursor;
14      //定义数据库名称及结构
15      static String TABLE_NAME = "Users";              //数据表名
16      static String ID = "_id";                        //ID
17      static String USER_NAME = "user_name";           //用户名
18      static String ADDRESS = "address";               //地址
```

```
19       static String TELEPHONE = "telephone";        //联系电话
20       static String MAIL_ADDRESS = "mail_address";   //电子邮箱
21     DBConnection(Context ctx)
22     {
23       super(ctx, Database_name, null, Database_Version);    ← 创建空数据库
24     }
25   public void onCreate(SQLiteDatabase database)
26   {
27     String sql = "CREATE TABLE " + TABLE_NAME + " ("
28         + ID + " INTEGER primary key autoincrement, "
29         + USER_NAME + " text not null, "               ← 创建数据表
30         + TELEPHONE + " text not null, "
31         + ADDRESS + " text not null, "
32         + MAIL_ADDRESS + " text not null "+ ");";
33     database.execSQL(sql);
34   }
35   public void onUpgrade(SQLiteDatabase db, int oldVersion, int newVersion)
36       {            }
37   }
```

(3) 设计控制程序 MainActivity.java。

```
1    package com.ex8_2;
2    import android.app.Activity;
3    import android.content.ContentValues;
4    import android.content.Context;
5    import android.database.Cursor;
6    import android.database.sqlite.SQLiteDatabase;
7    import android.os.Bundle;
8    import android.view.View;
9    import android.view.View.OnClickListener;
10   import android.widget.Button;
11   import android.widget.EditText;
12
13   public class MainActivity extends Activity
14   {
15     static EditText mEditText01;
16     static EditText mEditText02;
17     static EditText mEditText03;
18     static EditText mEditText04;
19     Cursor cursor;
20     Button createBtn, openBtn, upBtn, downBtn;
21     Button addBtn, updateBtn, deleteBtn, closeBtn;
22     SQLiteDatabase db;
23     DBConnection helper;
24     public int id_this;
```

```java
25      Bundle savedInstanceState;
26      //定义数据库名称及结构
27        static String TABLE_NAME = "Users";            //数据表名
28        static String ID = "_id";                      //ID
29        static String USER_NAME = "user_name";         //用户名
30        static String ADDRESS = "address";             //地址
31        static String TELEPHONE = "telephone";         //联系电话
32        static String MAIL_ADDRESS = "mail_address";   //电子邮箱
33      @Override
34      public void onCreate(Bundle savedInstanceState)
35      {
36         super.onCreate(savedInstanceState);
37         setContentView(R.layout.main);
38         mEditText01 = (EditText)findViewById(R.id.EditText01);
39         mEditText02 = (EditText)findViewById(R.id.EditText02);
40         mEditText03 = (EditText)findViewById(R.id.EditText03);
41         mEditText04 = (EditText)findViewById(R.id.EditText04);
42         createBtn = (Button)findViewById(R.id.createDatabase1);
43         createBtn.setOnClickListener(new ClickEvent());
44         openBtn = (Button)findViewById(R.id.openDatabase1);
45         openBtn.setOnClickListener(new ClickEvent());
46         upBtn=(Button)findViewById(R.id.up1);
47         upBtn.setOnClickListener(new ClickEvent());
48         downBtn=(Button)findViewById(R.id.down1);
49         downBtn.setOnClickListener(new ClickEvent());
50         addBtn = (Button)findViewById(R.id.add1);
51         addBtn.setOnClickListener(new ClickEvent());
52         updateBtn = (Button)findViewById(R.id.update1);
53         updateBtn.setOnClickListener(new ClickEvent());
54         deleteBtn = (Button)findViewById(R.id.delete1);
55         deleteBtn.setOnClickListener(new ClickEvent());
56         closeBtn = (Button)findViewById(R.id.clear1);
57         closeBtn.setOnClickListener(new ClickEvent());
58      }
59      class ClickEvent implements OnClickListener
60      {
61      public void onClick(View v)
62         {
63            switch(v.getId())
64            {
65            case R.id.createDatabase1:
66                helper = new DBConnection(MainActivity.this);
67                SQLiteDatabase db = helper.getWritableDatabase();
68                break;
69            case R.id.openDatabase1:
```

```java
70              db = openOrCreateDatabase("PhoneBook.db",
71                      Context.MODE_PRIVATE, null) ;           // 查询 Users 数据表
72              cursor = db.query("Users",
73                  null , null, null, null, null, null);
74              cursor.moveToNext();
75              upBtn.setClickable(true);
76              downBtn.setClickable(true);
77              deleteBtn.setClickable(true);
78              updateBtn.setClickable(true);
79              break;
80          case R.id.up1:
81           if(!cursor.isFirst())
82                  cursor.moveToPrevious();                     // 按下"上一记录"向前查询
83            datashow();
84             break;

85          case R.id.down1:
86             if(!cursor.isLast())
87                  cursor.moveToNext();                          // 按下"下一记录"向后查询
88             datashow();
89             break;
90          case R.id.add1:
91             add();                                             // 按下"添加"新增一行数据
92             onCreate(savedInstanceState);
93             break;
94          case R.id.update1:
95             update();                                          // 按下"修改"更新一行数据
96             onCreate(savedInstanceState);
97             break;
98          case R.id.delete1:
99             delete();                                          // 按下"删除"删除一行数据
100            onCreate(savedInstanceState);
101            break;
102         case R.id.clear1:                                     // 按下"关闭通讯录"关闭数据库
103            cursor.close();
104            mEditText01.setText("数据库已关闭");
105            mEditText02.setText("数据库已关闭");
106            mEditText03.setText("数据库已关闭");
107            mEditText04.setText("数据库已关闭");
108            upBtn.setClickable(false);
109            downBtn.setClickable(false);                       // 使按钮不可用
110            deleteBtn.setClickable(false);
111            updateBtn.setClickable(false);
112            break;
113         }
```

```
114     }
115   }
116   /* 显示记录 */
117   void datashow()
118   {
119     id_this = Integer.parseInt(cursor.getString(0));
120     String user_name_this = cursor.getString(1);
121     String telephone_this = cursor.getString(2);
122     String address_this = cursor.getString(3);
123     String mail_address_this = cursor.getString(4);
124     mEditText01.setText(user_name_this);
125     mEditText02.setText(telephone_this);
126     mEditText03.setText(address_this);
127     mEditText04.setText(mail_address_this);
128   }
129
130   /* 添加记录 */
131   void add()
132   {
133     ContentValues values1 = new ContentValues();
134     values1.put(USER_NAME, MainActivity.mEditText01.getText().toString());
135     values1.put(TELEPHONE, MainActivity.mEditText02.getText().toString());
136     values1.put(ADDRESS, MainActivity.mEditText03.getText().toString());
137     values1.put(MAIL_ADDRESS,
138             MainActivity.mEditText04.getText().toString());
139     SQLiteDatabase db2 = helper.getWritableDatabase();
140     db2.insert(TABLE_NAME, null, values1);
141     db2.close();
142   }
143   /* 修改记录 */
144   void update()
145   {
146     ContentValues values = new ContentValues();
147     values.put(USER_NAME, MainActivity.mEditText01.getText().toString());
148     values.put(TELEPHONE, MainActivity.mEditText02.getText().toString());
149     values.put(ADDRESS, MainActivity.mEditText03.getText().toString());
150     values.put(MAIL_ADDRESS,
151             MainActivity.mEditText04.getText().toString());
152     String where1 = ID + " = " + id_this;
```

119—123 读取各字段数据

140 将数据插入数据表

```
153    SQLiteDatabase db1 = helper.getWritableDatabase();
154    db1.update(TABLE_NAME, values, where1 ,null);   ← 替换数据表中的原数据
155    db1.close();
156  }
157  /* 删除记录 */
158  void delete()
159  {
160    String where = ID + " = " + id_this;
161    db.delete(TABLE_NAME, where ,null);   ← 删除数据表中的数据
162    db = helper.getWritableDatabase();
163    db.close();
164  }
165 }
```

程序的运行结果如图 8.5 所示。

图 8.5　数据库运行示例

8.2　文件的处理

8.2.1　输入/输出流

程序可以理解为数据输入、输出以及数据处理的过程，在程序执行过程中通常需要读取处理数据，并且将处理后的结果保存起来。Android 系统提供了对数据流进行输入与输出的方法。

1. 文件与目录管理的 File 类

Android 系统处理文件时直接调用 Java 语言的 java.io 包中的 File 类。每个 File 类的对

象都对应了系统的一个文件或目录,所以在创建 File 类对象时需要指明它所对应的文件或目录名。

下面是应用 File 类创建目录对象及文件对象的示例。

假设:

```
String sdir = "data/jtest";
String sfile = "FileIO.data";
```

则:

```
File Fdir = new File (sdir);              //目录对象
File Ffile = new File (Fdir, sfile);      //文件对象
```

一个对应于某文件或目录的 File 对象一经创建就可以通过调用它的方法来获得文件或目录的属性。在建立文件类的一个实例后可以查询这个文件对象,用测试方法获得文件或目录的有关信息,例如检测文件和目录的属性。File 类的常用方法见表 8-3。

表 8-3 File 类的常用方法

方法	说明
exists()	判断文件或目录是否存在
isFile()	判断对象是否为文件
isDirectory()	判断对象是否为目录
getName()	返回文件名或目录名
getPath()	返回文件或目录的路径
length()	返回文件的字节数
renameTo(File newFile)	将文件重命名成 newFile 对应的文件名
delete()	将当前文件删除
mkdir()	创建当前目录的子目录

2. 文件输入/输出流

在 Android 中处理二进制文件使用字节输入/输出流,处理字符文件使用字符输入/输出流。下面介绍对文件进行输入/输出处理的 4 个类。

- FileInputStream:字节文件输入流。
- FileOutputStream:字节文件输出流。
- FileReader:字符文件输入流。
- FileWriter:字符文件输出流。

8.2.2 处理文件流

1. 用文件输出流保存文件

1)FileOutputStream 类

FileOutputStream 类是从 OutputStream 类派生出来的输出类,它具有向文件中写数据的能力。它的构造方法有以下 3 种形式:

- FileOutputStream(String filename)
- FileOutputStream(File file)

- FileOutputStream(FileDescriptor fdObj)

其中各参数的含义如下。
- String filenam：指定的文件名，包括路径。
- File file：指定的文件对象。
- FileDescriptor fdObj：指定的文件描述符。

用户也可以通过 Context.openFileOutput()方法获取 FileOutputStream 对象。

2）把字节发送到文件输出流的 write()方法

输出流只是建立了一条通往数据要去的目的地的通道，数据并不会自动进入输出流通道，用户要使用文件输出流的 write()方法把字节发送到输出流。

使用 write()方法有 3 种格式。
- write(int b)：将指定字节写入此文件输出流。
- write(byte[] b)：将 b.length 个字节从指定字节数组写入此文件输出流中。
- write(byte[] b, int off, int len)：将指定字节数组中从偏移量 off 开始的 len 个字节写入此文件输出流。

【例 8-3】 把字符串"Hello World!"保存到本地资源的 test.txt 文件中。

在项目设计中设置一个"保存文件"按钮，其按钮事件调用下列方法：

```
1   void savefile()
2   {
3       String fileName="test.txt";
4       String str = "Hello World!";
5       FileOutputStream f_out;
6       try {
7           f_out = openFileOutput(fileName, Context.MODE_PRIVATE);  ← 文件输出流
8           f_out.write(str.getBytes());  ← 写操作
9       }
10      catch (FileNotFoundException e) {e.printStackTrace();}
11      catch (IOException e) {e.printStackTrace();}
12  }
```

则文件 test.txt 保存在××××data\data\（包名）\files 目录之下，应用 DDMS 工具可以查看到保存在本地资源目录下的文件，如图 8.6 所示。

2. 用文件输入流读取文件

1）FileInputStream 类

FileInputStream 类是从 InputStream 类中派生出来的输入流类，它用于处理二进制文件的输入操作。它的构造方法有下面 3 种形式：

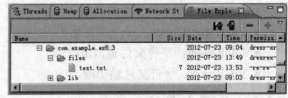

图 8.6 应用 DDMS 工具查看保存到本地资源目录下的文件

```
FileInputStream(String filename);
FileInputStream(File file);
FileInputStream(FileDescriptor fdObj);
```

其参数的含义和 FileInputStream 一样。

用户也可以通过 Context.openFileInput()方法获取 FileInputStream 对象。

2）从文件输入流中读取字节的 read()方法

文件输入流只是建立了一条通往数据的通道，应用程序可以通过这个通道读取数据，要实现读取数据的操作，需要使用 read()方法。

使用 read()方法有 3 种格式：
- int read();
- int read(byte b[ ]);
- int read(byte b[ ],int off, int len);

第 1 种格式每次只能从输入流中读取一个字节的数据。该方法返回的是一个 0～255 的整数值，若为文本类型的数据则返回的是 ASCII 值。如果该方法到达输入流的末尾则返回-1。

第 2 种格式和第 3 种格式以字节型数组作为参数，一次可以读取多个字节，读入的字节数据直接放入字节数组 b 中，并返回实际读取的字节个数。如果该方法到达输入流的末尾则返回-1。

第 3 种格式设置了偏移量（off）。这里的偏移量是指可以从字节型数组的第 off 个位置起读取 len 个数据。

【例 8-4】 读取本地资源文件 test.txt 中的内容。

在项目设计中设置一个"读取资源文件"按钮，其按钮事件调用下列方法：

```
1   void readfile()
2   {
3      String fileName="test.txt", str ;
4      byte[]  buffer = new byte[1024];    ← 设一字节数组，用于存放读取的数据
5      FileInputStream in_file=null;
6      try {
7         in_file = openFileInput(fileName);  ← 文件输入流
8         int  bytes = in_file.read(buffer);
9         str = new String(buffer, 0, bytes);  ← 将读取到的文件数据转换成字符串
10        Toast.makeText(MainActivity.this,
11           "文件内容：" + str, Toast.LENGTH_LONG).show();
12     }
13     catch (FileNotFoundException e) { System.out.print("文件不存在");}
14     catch (IOException e) { System.out.print("IO流错误");
15  }
```

3. 对 SD 卡文件的读/写

上述应用文件流对文件的读/写操作也适用于 SD 卡，但在处理上稍有不同，因为这里要考虑对 SD 卡的读/写权限。

1）环境变量访问类 Environment

Environment 为提供的环境变量访问类，在 Android 程序中对 SD 卡文件进行读/写操作时经常需要应用它的以下两个方法。

（1）getExternalStorageState()：获取当前存储设备的状态。

（2）getExternalStorageDirectory()：获取 SD 卡的根目录。

2）读/写 SD 卡的权限

（1）Environment 的主要常量为 Environment.MEDIA_MOUNTED，表示对 SD 卡具有

读/写权限。判断是否具有对 SD 卡文件进行读/写操作的权限通常使用下列条件语句:
if(Environment.getExternalStorageState().equals(Environment.MEDIA_MOUNTED))
（2）在 AndroidManifest.xml 文件中要加入允许对 SD 卡进行操作的权限语句。

- 允许在 SD 卡中创建及删除文件的权限语句：

```
<uses-permission
    android:name="android.permission.MOUNT_UNMOUNT_FILESYSTEMS">
</uses-permission>
```

- 允许往 SD 卡中写入数据的权限语句：

```
<uses-permission
    android:name="android.permission.WRITE_EXTERNAL_STORAGE">
</uses-permission>
```

【例 8-5】 读取与保存文件的应用程序示例。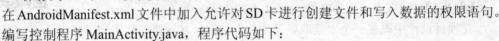
新建应用项目，设置一个文本编辑框和 4 个按钮，按钮分别为"保存到资源""保存到 SD 卡""读取资源文件"和"读取 SD 卡文件"。
在 AndroidManifest.xml 文件中加入允许对 SD 卡进行创建文件和写入数据的权限语句。
编写控制程序 MainActivity.java，程序代码如下：

```
1   package com.example.ex8_5;
2   import java.io.File;
3   import java.io.FileInputStream;
4   import java.io.FileNotFoundException;
5   import java.io.FileOutputStream;
6   import java.io.IOException;
7   import android.os.Bundle;
8   import android.os.Environment;
9   import android.view.View;
10  import android.view.View.OnClickListener;
11  import android.widget.Button;
12  import android.widget.EditText;
13  import android.widget.Toast;
14  import android.app.Activity;
15  import android.content.Context;
16  public class MainActivity extends Activity
17  {
18    Button saveBtn, readBtn, savesdBtn, readsdBtn;
19    EditText edit;
20    String fileName="test.txt";
21    String str;
22    @Override
23    public void onCreate(Bundle savedInstanceState)
24    {
25        super.onCreate(savedInstanceState);
26        setContentView(R.layout.activity_main);
```

```
27        edit=(EditText)findViewById(R.id.edit1);
28        saveBtn=(Button)findViewById(R.id.button1);
29        saveBtn.setOnClickListener(new mClick());
30        readBtn=(Button)findViewById(R.id.button2);
31        readBtn.setOnClickListener(new mClick());       ← 初始化组件
32        savesdBtn=(Button)findViewById(R.id.button3);
33        savesdBtn.setOnClickListener(new mClick());
34        readsdBtn=(Button)findViewById(R.id.button4);
35        readsdBtn.setOnClickListener(new mClick());
36    }
37    //按钮事件
38    class mClick implements OnClickListener
39    {
40        @Override
41        public void onClick(View arg0)
42        {
43            if(arg0== saveBtn)
44            {
45                savefile();      ← 写入文件
46            }
47            else if(arg0== readBtn)
48            {
49                readfile(fileName);   ← 读取资源文件数据
50            }
51            else if(arg0== savesdBtn)
52            {
53            setRegist();//动态获取本地存储卡（SD卡）读写权限
54                saveSDcar();    ← 写入到SD卡
55            }
56            else if(arg0 == readsdBtn)
57            {
58                readsdcard(fileName);  ← 读取SD卡文件数据
59            }
60        }
61    }
62
63    void savefile()
64    {
65      str= edit.getText().toString();
66      try {
67            FileOutputStream f_out =                      ← 创建文件
68            openFileOutput(fileName, Context.MODE_PRIVATE);  输出流
69            f_out.write(str.getBytes());  ← 文件输出流按字节输出数据到文件中
70        }catch (FileNotFoundException e) {
71            e.printStackTrace();
72        } catch (IOException e) {
73            e.printStackTrace();
74        }
75    }
76    //读取文件数据
```

```java
77  void readfile(String fileName)
78  {
79      byte[] buffer = new byte[1024];         // 创建字节数组存放数据
80      FileInputStream in_file=null;
81      try {
82          in_file = openFileInput(fileName);      // 文件输入流将读到的数据放
83          int  bytes = in_file.read(buffer);      // 入字节数组,再将字节转换成
84          str = new String(buffer, 0, bytes);     // 字符串
85          Toast.makeText(MainActivity.this,
86              "文件内容: " + str, Toast.LENGTH_LONG).show();
87      } catch (FileNotFoundException e) { System.out.print("文件不存在");}
88      catch (IOException e) { System.out.print("IO流错误");}
89  }
90  //保存文件到SD卡
91  void saveSDcar()
92  {
93    str= edit.getText().toString();
94    if(Environment.getExternalStorageState()         // 判断SD卡是否允许读/写操作
95        .equals(Environment.MEDIA_MOUNTED))
96    {
97    File path = Environment.getExternalStorageDirectory();  // 获取SD卡目录路径
98    File sdfile = new File(path, fileName);
99    try {
100         FileOutputStream f_out = new FileOutputStream(sdfile);
101         f_out.write(str.getBytes());           // 文件输出流将数据写入SD卡
102         Toast.makeText(MainActivity.this,
103             "文件保存到SD卡", Toast.LENGTH_LONG).show();
104     }catch (FileNotFoundException e)
105     {
106         e.printStackTrace();
107     } catch (IOException e) {
108         e.printStackTrace();
109     }
110   }
111 }
112 //从SD卡读取文件内容
113 void readsdcard(String fileName)
114 {
115   if(Environment.getExternalStorageState()
116       .equals(Environment.MEDIA_MOUNTED))       // 判断SD卡是否允许读/写操作
117   {
118     File path = Environment
119         .getExternalStorageDirectory();         // 获取SD卡目录路径
120     File sdfile= new File(path, fileName);
121     try {
122         FileInputStream in_file =new FileInputStream(sdfile);
123         byte[] buffer = new byte[1024];
124         int  bytes = in_file.read(buffer);      // 读取文件数据到字节数组
125         str = new String(buffer, 0, bytes);
126         Toast.makeText(MainActivity.this,
```

```
127                 "文件内容: " + str, Toast.LENGTH_LONG).show();
128             } catch (FileNotFoundException e){ System.out.print("文件不存在");}
129             catch (IOException e) { System.out.print("IO流错误");}
130         }
131     }
132     private void setRegist() {
        /*
         * 动态获取权限
         * Android 6.0 新特性, 一些保护权限, 如, 文件读写除了要在
         * AndroidManifest中声明权限, 还要使用如下代码动态获取
         */
133         if (Build.VERSION.SDK_INT >= 23) {//大于23是指Android 6.0以后版本
134             int REQUEST_CODE_CONTACT = 101;
135             String[] PERMISSIONS_STORAGE = {
136                     Manifest.permission.READ_EXTERNAL_STORAGE,
137                     Manifest.permission.WRITE_EXTERNAL_STORAGE};
            //验证是否许可权限
138             for (String str : PERMISSIONS_STORAGE) {
139                 if (this.checkSelfPermission(str) !=
140                         PackageManager.PERMISSION_GRANTED ) {
                    //申请权限
141                     this.requestPermissions(PERMISSIONS_STORAGE,
142                             REQUEST_CODE_CONTACT);
143                     return;
144                 }
145             }
146         }
147     }
```

程序的运行结果如图 8.7 所示。

图 8.7　读/写文件示例

8.3 轻量级存储 SharedPreferences

Android 系统提供了一个存储少量数据的轻量级的数据存储方式 SharedPreferences。该存储方式类似于 Web 程序中的 Cookie，通常用它来保存一些配置文件数据、用户名及密码等。SharedPreferences 采用"键-值"对的形式组织和管理数据，其数据存储在 XML 格式的文件中。

使用 SharedPreferences 方式存储数据需要用到 SharedPreferences 和 SharedPreferences.Editor 接口，这两个接口在 android.content 包中。

SharedPreferences 对象由 Context.getSharedPreferences（String name，int mode）方法构造，它有两个参数，其含义如下。

（1）第 1 个参数 name 为保存数据的文件名，因为 SharedPreferences 是使用 XML 文件保存数据，getSharedPreferences(name,mode)方法的第 1 个参数用于指定该文件的名称，名称不用带后缀，后缀会由 Android 自动加上。该 XML 文件存放在 data\data\××××（包名）\shared_prefs 目录下。

（2）第 2 个参数 mode 为操作模式。

- MODE_PRIVATE：这是默认的形式，配置文件只允许本程序和享有本程序 ID 的程序访问。
- MODE_WORLD_READABLE：允许其他的应用程序读文件。
- MODE_WORLD_WRITEABLE：允许其他的应用程序写文件。
- MODE_MULTI_PROCESS：主要用于多任务，当多个进程共同访问的时候必须指定这个标签。

SharedPreferences 对象的常用方法见表 8-4。

表 8-4　SharedPreferences 接口的常用方法

方法	说明
edit()	建立一个 SharedPreferences.Editor 对象
contains(String key)	判断是否包含该键值
getAll()	返回所有配置信息
getBoolean(String key,boolean defValue)	获得一个 boolean 类型的数据
getFloat(String key,float defValue)	获得一个 float 类型的数据
getInt(String key,int defValue)	获得一个 int 类型的数据
getLong(String key,long defValue)	获得一个 long 类型的数据
getString(String key,String defValue)	获得一个 String 类型的数据

SharedPreferences.Editor 接口用于存储 SharedPreference 对象的数据值，SharedPreferences.Editor 接口的常用方法见表 8-5。

表 8-5　SharedPreferences.Editor 接口的常用方法

方法	说明
clear()	清除所有数据值
commit()	保存数据

续表

方法	说明
putBoolean（String key, boolean value）	保存一个 boolean 类型的数据
putFloat（String key, float value）	保存一个 float 类型的数据
putInt（String key, int value）	保存一个 int 类型的数据
putLong（String key, long value）	保存一个 long 类型的数据
putString（String key, String value）	保存一个 String 类型的数据
remove（String key）	删除键名 key 所对应的数据值

SharedPreference.Editor 的 putXXX 方法以键-值对的形式存储数据，最后一定要调用 commit 方法提交数据文件才能保存。

读取数据非常简单，直接调用 SharedPreference 对象相应的 get×××方法即可获得数据。

【例 8-6】 应用 SharedPreferences 对象将一个客户的联系电话保存到电话簿中。

设客户名为 zsm，其电话为 123456。故设电话簿的文件名为 phoneBook，其数据的"键-值"对为（"name", "zsm"）和（"phone", "123456"）。

程序设计如下：

```
1   package com.example.ex8_6;
2   import android.os.Bundle;
3   import android.app.Activity;
4   import android.content.Context;
5   import android.content.SharedPreferences;
6   import android.view.View;
7   import android.view.View.OnClickListener;
8   import android.widget.Button;
9   import android.widget.Toast;
10
11    public class MainActivity extends Activity
12    {
13       SharedPreferences settings;
14  Button saveBtn;
15       @Override
16       public void onCreate(Bundle savedInstanceState)
17       {
18         super.onCreate(savedInstanceState);
19         setContentView(R.layout.activity_main);
20         saveBtn=(Button)findViewById(R.id.button1);
21         saveBtn.setOnClickListener(new mClick());
22       }
23       //按钮事件
24       class mClick implements OnClickListener
25       {
26  public void onClick(View arg0)
27       {
```

```
28              settings = getSharedPreferences("phoneBook", Context.MODE_
                PRIVATE);
29              SharedPreferences.Editor editor = settings.edit();
30              editor.putString("name", "zsm");
31              editor.putString("phone", "123456");
32              editor.commit();
33          Toast.makeText(MainActivity.this,"保存成功!",Toast.LENGTH_LONG);
34      }
35   }
36 }
```

数据文件 phoneBook.xml 保存在 data\data\com.example.ex8_6（包名）\shared_prefs 目录之下，扩展名.xml 由系统自动生成。应用 DDMS 工具可以查看到该文件，如图 8.8 所示。

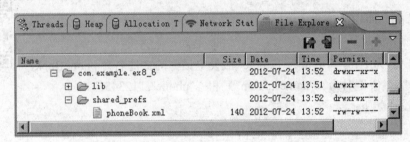

图 8.8　应用 DDMS 工具查看 XML 文件

单击 DDMS 工具的 按钮可以将数据文件 phoneBook.xml 从模拟器中保存到计算机的磁盘中，打开该 XML 格式的数据文件 phoneBook.xml，其内容如下：

```
<?xml version='1.0' encoding='UTF-8' standalone='yes' ?>
<map>
<string name="phone">123456</string>
<string name="name">zsm</string>
</map>
```

8.4　访问远程数据库

Android 访问远程数据库，通常的方法是访问远程服务器，由服务器程序连接后台数据库。

下面通过一个示例来说明访问远程数据库的方法。

【例 8-7】　编写一个访问远程数据库的应用程序。

（1）设有 MySQL 数据库，数据库名为 testdb。该数据库中有数据表 user，其数据结构如表 8-6 所示。

表 8-6　数据表 user 的数据结构

字段名	字段类型	说明
sid	int(5)	序号

续表

字段名	字段类型	说明
name	varchar(10)	姓名
email	varchar(25)	邮箱

(2) 界面布局设计。在界面中设置一个按钮 Button 和一个列表组件 ListView。

(3) 把 volley.jar 复制到项目的 app\libs 目录下，并完成 jar 包的安装。

(4) 设计主类，其代码如下：

```
1   package com.example.ex8_Mysql_volley;
2   import android.support.v7.app.AppCompatActivity;
3   import android.os.Bundle;
4   import android.util.Log;
5   import android.view.View;
6   import android.widget.ArrayAdapter;
7   import android.widget.Button;
8   import android.widget.ListView;
9   import android.widget.TextView;
10  import com.android.volley.RequestQueue;
11  import com.android.volley.Response;
12  import com.android.volley.VolleyError;
13  import com.android.volley.toolbox.StringRequest;
14  import com.android.volley.toolbox.Volley;
15  import org.json.JSONArray;
16  import org.json.JSONException;
17  import org.json.JSONObject;
18  public class MainActivity extends AppCompatActivity {
19    TextView txt;
20    Button volleyBtn;
21    ListView listdata;
22    Datainfo datainfo=new Datainfo();
23    @Override
24    protected void onCreate(Bundle savedInstanceState) {
25      super.onCreate(savedInstanceState);
26      setContentView(R.layout.activity_main);
27      volleyBtn=(Button)findViewById(R.id.button);
28      txt = (TextView) findViewById(R.id.textView);
29      listdata = (ListView)findViewById(R.id.listView);
30      getServerData();
31      volleyBtn.setOnClickListener(new mClick());
32    }
33    public class Datainfo     //该类用于数据的输入和输出
34    {
35      private String sid;     //序号
```

```java
36    private String name;    //姓名
37    private String email;   //邮箱
38    public void setSid(String sid){this.sid = sid;}
39    public void setName(String name){this.name = name;}
40    public void setEmail(String email){this.email = email;}
41    public String getName(){return name;}
42    public String getSid(){return sid;}
43    public String getEmail(){return email;}
44  }
45  class mClick implements View.OnClickListener
46  {
47    String[] list={"","",""};
48    @Override
49    public void onClick(View v) {
50      list[0] = "序号：" + datainfo.getSid();
51      list[1] = "姓名：" + datainfo.getName();
52      list[2] = "邮箱：" + datainfo.getEmail();
53      ArrayAdapter<String> adapter = new ArrayAdapter<String>(
54        MainActivity.this,
55        android.R.layout.simple_list_item_1,
56        list);
57      listdata.setAdapter(adapter);
58    } //onClick()_end
59  } //class mClick_end

60  public void getServerData(){
61    String jsonURL = "http://58.199.89.161/test/conn_zsmdb.php";
62    try {
63      RequestQueue mQueue = Volley.newRequestQueue(MainActivity.this);
64      StringRequest stringRequest = new StringRequest(
65        jsonURL,//第1个参数，请求的网址
66        new Response.Listener<String>() { //第2个参数，响应正确时的处理
67          @Override
68          public void onResponse(String response) {
69            try {
70              JSONArray jsonArray = new JSONArray(response);
71              for (int i = 0; i < jsonArray.length(); i++) {
72                JSONObject jsonData= jsonArray.getJSONObject(i);
73                String sid = (String)jsonData.get("sid");
74                String sname = (String)jsonData.get("name");
75                String semail = (String)jsonData.get("email");
76                datainfo.setSid(sid);
77                datainfo.setName(sname);
78                datainfo.setEmail(semail);
79              }
```

```
80            }catch (JSONException e) {
81              txt.setText("list错误");
82              Log.e("json错误", e.getMessage(), e);
83            }
84          }
85        },
86        new Response.ErrorListener() {    //第3个参数，错误时反馈信息
87          @Override
88          public void onErrorResponse(VolleyError error) {
89              Log.e("TAG错误", error.getMessage(), error);
90          }
91      });  //StringRequest()_end
92      mQueue.add(stringRequest);
93      }catch (Exception e){  }
94    } //getServerData()_end
95  } //class MainActivity_end
```

(5) 修改配置文件，添加访问网络权限：

```
<uses-permission android:name="android.permission.INTERNET" />
```

(6) 服务器端连接数据库的 PHP 文件 conn_testdb.php。

```
1   <?php
2     header("Content-Type: text/html;charset=utf8");
                                                    //设置页面显示的文字编码
3     $con = mysql_connect("localhost", "root", ""); //连接数据库
4     mysql_query("set names utf8");//设置数据库的编码方式
5     //下面一大段代码是为了拼接出JSON格式的字符串
6     echo "var json = [";
7     if($con)
8     {
9       mysql_select_db("testdb", $con);          //选择要使用的数据库
10      $query = "SELECT * FROM user";            //数据库查询语句
11      $result = mysql_query($query);            //执行查询操作
12      $i = 0;                                    //用来判断是否为第1条数据
13      while($row = mysql_fetch_array($result))
14      {
15        if($i != 0){ echo ","; }//如果是第1条数据，则在数据前不显示逗号分隔符
16        else{ $i = 1;   }
17        echo '{ "';
18        echo 'sid":';
19        echo '"';
20        echo $row['sid'];
21        echo '",';
22        echo '"';
```

```
23      echo 'name":';
24      echo '"';
25      echo $row['name'];
26      echo '",';
27      echo '"';
28      echo 'email":';
29      echo '"';
30      echo $row['email'];
31      echo '"}';
32    }
33  }else
34  {
35    //如果连接数据库失败,仍然可以返回一条JSON数据
36    echo '{ "sid":101,"name":"服务器出错了","email":"123@abc"}';
37  }
38  echo "]";
39  mysql_close($con);
40  ?>
```

程序的运行结果如图 8.9 所示。

图 8.9 访问远程数据库

习 题 8

1. 编写一个小型商场的销售管理系统，可以输入商品的名称、数量、单价，并具有汇总功能。

2. 编写一个如图 8.10 所示的小型记事本，以文件形式保存。

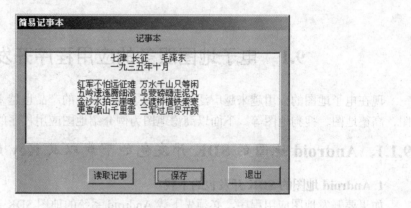

图 8.10 小型记事本

3. 设计一个远程访问数据库的应用程序，实现具有密码验证的用户登录功能。

第 9 章　地图服务与传感器检测技术

9.1　电子地图服务的应用程序开发

现在电子地图的应用越来越广泛，提供电子地图服务的产品也越来越多，例如百度地图、高德地图、搜狗地图等。下面以高德地图为例介绍地图应用程序的开发方法。

9.1.1　Android 地图的 SDK 开发包的下载以及 Key 的申请

1. Android 地图的 SDK 开发包的下载

如果要开发地图应用程序，必须先下载 Android 系统的地图 SDK 开发包。

高德地图的 Android SDK 是一套地图开发调用接口，开发者可以很方便地在 Android 应用中加入地图的相关功能，包括地图的显示（含室内、室外地图）、与地图交互、在地图上绘制、兴趣点搜索、地理编码、离线地图等功能。

高德地图的 Android 地图 SDK 的下载地址如下：

http://lbs.amap.com/api/android-sdk/download

如图 9.1 所示。

图 9.1　下载 Android 地图 SDK

2. 申请地图服务 Key

在进行地图服务的应用程序开发之前必须申请一组验证过的 Key，这样才可以使用地图服务。

如果想获得高德地图的 Key，需要提供两个信息——SHA1 和地图应用程序的包名。

1）获取 SHA1

打开 Android ADK 所在目录，通常在 C 盘的用户目录中以.android 为文件夹名。该文件夹中有名为"default.keystore"的文件，如图 9.2 所示。

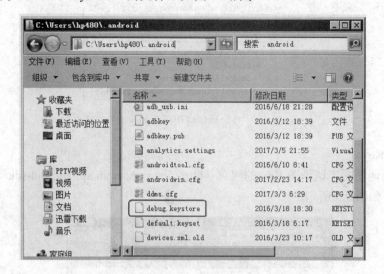

图 9.2　打开 Android ADK 所在目录

这个 debug.keystore 文件就是获取 SHA1 值的文件。接下来按 WIN+R 键，输入 cmd，打开命令行窗口，进入到.android 文件夹中，输入以下命令：

keytool -list -v -keystore debug.keystore

系统提示输入密钥库密码，开发模式的默认密码是 android。输入密钥后回车，此时可在控制台显示的信息中获取到证书指纹 SHA1 的值，如图 9.3 所示（说明：keystore 文件为 Android 签名证书文件）。

图 9.3　得到证书指纹 SHA1 的值

2）获取地图应用项目的包名

新建地图应用项目，打开 Android 项目的 AndroidManifest.xml 配置文件，package 属性所对应的内容为应用包名，如图 9.4 所示。

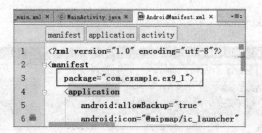

图 9.4 获取地图应用项目的包名

3）获取 Key

打开高德地图的申请地图 Key 的网页（http://lbs.amap.com/api/android-sdk/）选择"获取 Key"项，如图 9.5 所示。

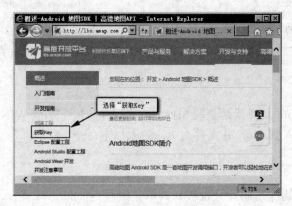

图 9.5 选择"获取 Key"

在弹出的设置 Key 的对话框中输入 SHA1 值和包名，单击"提交"按钮，如图 9.6 所示。

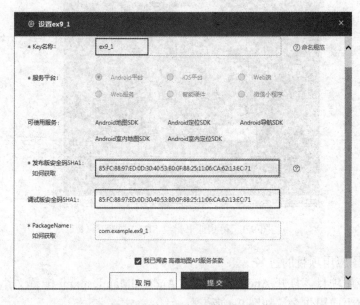

图 9.6 输入 SHA1 值和包名

单击"提交"按钮后得到应用程序的 Key，如图 9.7 所示。

Key名称	Key	绑定服务
ex9_1	d96f8e900571c7f9db51a89829a367f5	Android平台

图 9.7　得到地图应用程序的 Key

9.1.2　显示地图的应用程序示例

【例 9-1】　创建一个显示地图的应用程序。

创建名称为 ex9_1 的新项目，包名为 com.example.ex9_1。

（1）导入地图开发包。将下载的地图开发包 Amap2DMap_4.2.0.jar 导入到应用项目的 app\libs 目录下。然后右击开发包 Amap2_DMap_V4.2.0.jar，在弹出的菜单中选择 Add As Library 命令完成 jar 包的安装，如图 9.8 所示。

图 9.8　导入并安装地图开发包到 app\libs 目录下

（2）界面布局文件设计。在界面布局文件 activity_main.xml 中添加地图视图组件 MapView：

```xml
<com.amap.api.maps2d.MapView
    android:layout_width="match_parent"
    android:layout_height="match_parent"
    android:id="@+id/map">
</com.amap.api.maps2d.MapView>
```

界面布局文件的完整代码如下：

```
1   <?xml version="1.0" encoding="utf-8"?>
2   <RelativeLayout xmlns:android="http://schemas.android.com/apk/res/
    android"
3       xmlns:tools="http://schemas.android.com/tools"
4       android:layout_width="match_parent"
5       android:layout_height="match_parent"
6       android:paddingBottom="@dimen/activity_vertical_margin"
```

```
7      android:paddingLeft="@dimen/activity_horizontal_margin"
8      android:paddingRight="@dimen/activity_horizontal_margin"
9      android:paddingTop="@dimen/activity_vertical_margin"
10     tools:context="com.example.administrator.ex9_1.MainActivity">
11
12     <com.amap.api.maps2d.MapView
13         android:layout_width="match_parent"              ← 添加地图显示组
14         android:layout_height="match_parent"
15         android:id="@+id/map">
16     </com.amap.api.maps2d.MapView>
17 </RelativeLayout>
```

（3）设计主控程序。

```
1  package com.example.ex9_1;
2  import android.support.v7.app.AppCompatActivity;
3  import android.os.Bundle;
4  import com.amap.api.maps2d.AMap;
5  import com.amap.api.maps2d.MapView;
6  public class MainActivity extends AppCompatActivity {
7    @Override
8    protected void onCreate(Bundle savedInstanceState) {
9        super.onCreate(savedInstanceState);
10       setContentView(R.layout.activity_main);
11       MapView mapView = (MapView) findViewById(R.id.map);
12       mapView.onCreate(savedInstanceState);    //调用onCreate()方法
13       AMap aMap = mapView.getMap();
14       aMap.setMapType(AMap.MAP_TYPE_NORMAL);   //标准地图模式
15    }
16 }
```

说明：地图 SDK 开发包提供了 3 种地图模式，即 MAP_TYPE_NORMAL、MAP_TYPE_SATELLITE 和 MAP_TYPE_NIGHT。

- MAP_TYPE_NORMAL：标准地图。地图包含道路、建筑以及重要的自然风光（如河流）等。道路和功能标签为可见。
- MAP_TYPE_SATELLITE：卫星地图。3D 地图道路和功能标签为可见，2D 地图道路和功能标签不可见。
- MAP_TYPE_NIGHT：夜景地图（仅 3D 地图）。道路和功能标签可见。

（4）修改配置文件，设置地图 Key 及访问权限。

```
1  <?xml version="1.0" encoding="utf-8"?>
2  <manifest xmlns:android="http://schemas.android.com/apk/res/android"
3   package="com.example.ex9_1">
4   <application
5      android:allowBackup="true"
6      android:icon="@mipmap/ic_launcher"
7      android:label="@string/app_name"
8      android:supportsRtl="true"
```

```
9       android:theme="@style/AppTheme">
10      <activity android:name=".MainActivity">
11        <meta-data
12            android:name="com.amap.api.v2.apikey"
13            android:value="d96f8e900571c7f9db51a89829a367f5">    ◄── 地图的 Key
14        </meta-data>
15        <intent-filter>
16          <action android:name="android.intent.action.MAIN" />
17          <category android:name="android.intent.category.LAUNCHER" />
18        </intent-filter>
19      </activity>
20    </application>
21    <!--地图SDK（包含其搜索功能）需要的基础权限-->
22    <!--允许程序打开网络套接字-->
23    <uses-permission android:name="android.permission.INTERNET" />
24    <!--允许程序设置内置SD卡的写权限-->
25    <uses-permission android:name="android.permission.WRITE_EXTERNAL_
      STORAGE" />
26    <!--允许程序获取网络状态-->
27    <uses-permission android:name="android.permission.ACCESS_NETWORK_STATE"
      />
28    <!--允许程序访问WiFi网络信息-->
29    <uses-permission android:name="android.permission.ACCESS_WIFI_STATE" />
30    <!--允许程序读/写手机状态和身份-->
31    <uses-permission android:name="android.permission.READ_PHONE_STATE" />
32    <!--允许程序通过访问CellID或WiFi热点来获取粗略的位置-->
33    <uses-permission android:name="android.permission.ACCESS_COARSE_LOCATION"
      />
34
35    </manifest>
```

程序的运行结果如图9.9所示。

图 9.9　显示地图

9.2 传感器检测技术

9.2.1 传感器简介

传感器(sensor)是一种检测装置,它能检测和感受到外界的信号,并将信息变换成电信号或其他所需形式的信息输出,以满足信息的传输、处理、存储、显示、记录和控制等要求。

1. 传感器的类型

Android 系统中内置了很多类型的传感器,这些传感器被封装在 Sensor 类中。Sensor 类是管理各种传感器的共同属性(名字、版本等)的类,Sensor 类包含了一个常量集合,用于描述 Sensor 对象所表示的硬件传感器类型,这些常量均以 Sensor.TYPE_<TYPE>的形式表示。Android 系统的常见传感器类型见表 9-1。

表 9-1 Android 的常见传感器类型

类型常量	说明
Sensor.TYPE_ACCELEROMETER	加速度(重力)传感器
Sensor.TYPE_LIGHT	光线传感器
Sensor.TYPE_MAGNETIC_FIELD	磁场传感器
Sensor.TYPE_PROXIMITY	距离(临近性)传感器
Sensor.TYPE_AMBIENT_TEMPERATURE	温度传感器
Sensor.TYPE_PRESSURE	压力传感器
Sensor.TYPE_ALL	所有类型的传感器

2. 与传感器相关的类

除 Sensor 类之外,要使用传感器,还需要用到传感器管理类 SensorManager 和传感器事件监听接口 SensorEventListener。

1)传感器管理类 SensorMannager

Android 中的所有传感器都需要通过 SensorMannager 对象来访问,SensorMannager 没有构造方法,需要调用 getSystemService(SENSOR_SERVICE)方法创建传感器管理对象。

```
SensorManager mSensorMgr = (SensorManager) getSystemService(SENSOR_SERVICE);
```

SensorManager 类的常用方法见表 9-2。

表 9-2 传感器管理类 **SensorManager** 的常用方法

方法	说明
getSensorList(int type)	获取传感器类型列表
registerListener(　　SensorEventListener listener, 　　Sensor sensor, 　　int rate)	注册传感器的监听器
unregisterListener(SensorEventListener listener)	注销传感器的监听器
getDefaultSensor(int type)	获取默认的传感器对象

在 registerListener()方法中，第 3 个参数 rate 为传感器的更新速率，其更新速率分为 4 个级别。
- SensorManager.SENSOR_DELAY_FASTEST：最快级，特别敏感，一般不推荐使用。
- SensorManager.SENSOR_DELAY_GAME：游戏级，实时性较高的游戏使用。
- SensorManager.SENSOR_DELAY_NORMAL：普通级，默认使用。
- SensorManager.SENSOR_DELAY_UI：用户界面级，一般屏幕更新使用。

2）实现 SensorEventListener 接口

传感器事件监听接口 SensorEventListener 有两个方法必须实现。
- onAccuracyChanged(Sensor sensor,int accuracy)：传感器的精度变化的时候此方法被调用。
- onSensorChanged(SensorEvent event)：传感器的值改变的时候此方法被调用。

【例 9-2】 列出传感器的类型。

```
1   package com.example.sensortest;
2   import java.util.List;
3   import android.hardware.Sensor;
4   import android.hardware.SensorEvent;
5   import android.hardware.SensorEventListener;
6   import android.hardware.SensorManager;
7   import android.os.Bundle;
8   import android.widget.LinearLayout;
9   import android.widget.TextView;
10  import android.app.Activity;
11
12  public class MainActivity extends Activity
13   implements SensorEventListener
14  {
15    private SensorManager sensorManager;
16    @Override
17    public void onCreate(Bundle savedInstanceState)
18    {
19       super.onCreate(savedInstanceState);
20       setContentView(R.layout.activity_main);
21    }
22    @Override
23    protected void onResume()
24    {
25       super.onResume();
26       sensorManager =
27         (SensorManager)this.getSystemService(SENSOR_SERVICE);  ←获取传感器管理服务
28       List<Sensor> sensorList =
29         sensorManager.getSensorList(Sensor.TYPE_ALL);
```

```
30      LinearLayout layout = new LinearLayout(this);    ← 创建布局对象
31      layout.setOrientation(LinearLayout.VERTICAL);
32      TextView txt;
33      for (Sensor s:sensorList)
34      {
35          txt = new TextView(this);
36          txt.setText(s.getName());
37          layout.addView(txt,new LinearLayout.LayoutParams(
38                  LinearLayout.LayoutParams.FILL_PARENT,      设置
39                  LinearLayout.LayoutParams.WRAP_CONTENT));    布局
40      }
41      setContentView(layout);
42  }
43  @Override
44  public void onAccuracyChanged(Sensor sensor, int accuracy)
45  {            }
46  @Override
47  public void onSensorChanged(SensorEvent event)
48  {            }
49  }
```

在真实手机上运行程序列出传感器的类型,如图 9.10 所示。

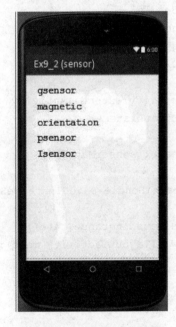

图 9.10　列出传感器的类型

9.2.2　加速度传感器的应用示例

加速度传感器是用于检测物体的加速度。物体在运动时其加速度也跟着变化,如果能

获取到加速度的值就可以知道物体受到什么样的作用力或物体进行什么样的运动。

通过 Android 的加速度传感器可以从 X、Y、Z 3 个方向轴获取加速度。X、Y、Z 3 个方向轴的定义如下：

- X 轴的方向是沿着手机屏幕从左向右的方向；
- Y 轴的方向是从手机屏幕的左下角开始沿着屏幕的上下方向指向屏幕的顶端；
- Z 轴的方向是从手机里指向外的前后方向。

加速度传感器的方向轴如图 9.11 所示。

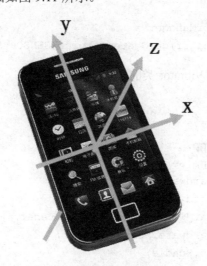

图 9.11 加速度传感器的方向轴

通过 SensorEventListener 接口的 onSensorChanged(SensorEvent event)可以获取 X、Y、Z 3 个方向轴重力加速度的值。

例如设 X 轴、Y 轴、Z 轴 3 个方向的重力分量的值分别为 x、y、z，则

```
public void onSensorChanged(SensorEvent event)
{
    float x = event.values[0];
    float y = event.values[1];
    float z = event.values[2];
}
```

【例 9-3】 将手机设置成振动状态，"摇一摇"后立刻停止振动。

本例要解决两个问题，一是将机器设为振动状态；二是应用加速度传感器的工作原理，快速晃动机器，即"摇一摇"后使机器振动停止。

Android 手机的振动控制由 Vibrator 类实现。Vibrator 类主要有下面两个方法。

（1）vibrate(long[] pattern, int repeat)方法：设置振动周期。

（2）cancel()方法：停止振动。

设置振动事件需要知道其振动的时间长短、振动的周期等。在 Android 中振动的时间以毫秒为单位（1/1000 秒）。注意，如果设置的时间值太小会感觉不出来。

通过调用 Vibrator 类的 vibrate(long[] pattern, int repeat)方法实现振动功能。vibrate()有两个参数，其含义如下。

- 第 1 个参数 long[] pattern：设置振动的效果的数组，数组中数字的含义依次是静止时长、振动时长、静止时长、振动时长，时长的单位是毫秒。
- 第 2 个参数 repeat：取值为-1 表示只振动一次，取值为 0 则振动会一直持续。

需要指出的是 Vibrator 类没有构造方法，需要通过建立对象引用的 getSystemService( )方法获取 Vibrator 对象：

```
SensorManager mSensorManager =
        (SensorManager)getSystemService(SENSOR_SERVICE);
```

若要停止振动，则调用 Vibrator 类的 cancel()方法即可。
程序代码如下：

```
1    package com.ex9_3;
2    import android.app.Activity;
3    import android.app.Service;
4    import android.hardware.Sensor;
5    import android.hardware.SensorEvent;
6    import android.hardware.SensorEventListener;
7    import android.hardware.SensorManager;
8    import android.os.Bundle;
9    import android.os.Vibrator;
10   import android.view.View;
11   import android.view.View.OnClickListener;
12   import android.widget.Button;
13
14      public class MainActivity extends Activity implements SensorEventListener
15   {
16      Button clearBtn;
17      private SensorManager mSensorManager;
18      private Vibrator vibrator;          ◄── 声明振动对象
19      @Override
20      public void onCreate(Bundle savedInstanceState)
21      {
22         super.onCreate(savedInstanceState);
23         setContentView(R.layout.main);
24         mSensorManager=
25           (SensorManager)getSystemService(SENSOR_SERVICE);   ◄── 获取传感器管理服务
26         vibrator = (Vibrator)getApplication()
27                .getSystemService(Service.VIBRATOR_SERVICE);  ◄── 获取振动对象
28         clearBtn = (Button) findViewById(R.id.clear);
29         clearBtn.setOnClickListener(new mClick());
30      }
```

```
31
32  class mClick implements OnClickListener
33  {
34      public void onClick(View v)
35      {
36      clearBtn.setText("改振动喽~");
37      try{
38          vibrator.vibrate(new long[]{100, 100, 100, 1000}, 0);   ← 设置振动周期
39          }catch(Exception e){
40              System.out.println("振动错误!!!!!! ");
41          }
42      }
43   }
44  @Override
45  protected void onResume()
46  {
47      super.onResume();
48      mSensorManager.registerListener(this,   ← 为加速度传感器注册监听器
49          mSensorManager.getDefaultSensor(Sensor.TYPE_ACCELEROMETER),
50          SensorManager.SENSOR_DELAY_NORMAL );
51  }
52  @Override
53  protected void onStop()
54  {
55      super.onStop();
56  }
57  @Override
58  protected void onPause()
59  {
60      super.onPause();
61  }
62  public void onAccuracyChanged(Sensor sensor, int accuracy)
63  {            }
64  public void onSensorChanged(SensorEvent event)   ← 当传感器的值发生改变时回调该方法
65  {
66      int sensorType = event.sensor.getType();
67      float[] values = event.values;    ← values[0]: X轴, values[1]: Y轴, values[2]: Z轴
68      if(sensorType == Sensor.TYPE_ACCELEROMETER )
69      {
70          /* 由于一般情况下任意轴数值在9.8~10,
71           * 当突然摇动手机的时候瞬时加速度突然增大或减少。
72           * 所以只需监听任意轴的加速度大于14的时候改变需要的设置
73           */
74          if((Math.abs(values[0])>14 || Math.abs(values[1])>14
75                  || Math.abs(values[2])>14))
```

```
76          {
77              clearBtn.setText("别摇了,头晕死了,已经停止振动");
78              vibrator.cancel();    ← 摇动手机后停止振动
79          }
80      }
81  }
82  public void onAccuracyChanged(int sensor, int accuracy)
83  {    }
84  public void onSensorChanged(int sensor, float[] values)
85  {    }
86 }
```

修改配置文件 AndroidManifest.xml,设置允许使用振动效果的权限:

```
<uses-permission android: name = "android.permission.VIBRATE"/>
```

由于模拟器无法实现振动功能,只能在真实手机上运行。程序运行如图 9.12 所示。当单击"设为振动"按钮后手机产生振动,快速摇动手机则振动停止。

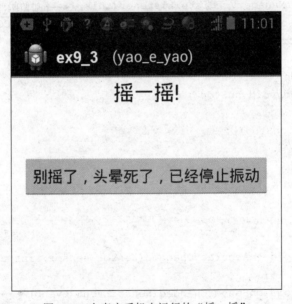

图 9.12 在真实手机上运行的"摇一摇"

【例 9-4】 简单的重力小球游戏。

```
1   package com.example.ex9_4;
2   import android.app.Activity;
3   import android.content.Context;
4   import android.content.pm.ActivityInfo;
5   import android.graphics.Bitmap;
6   import android.graphics.BitmapFactory;
7   import android.graphics.Canvas;
8   import android.graphics.Color;
```

```java
9    import android.graphics.Paint;
10   import android.hardware.Sensor;
11   import android.hardware.SensorEvent;
12   import android.hardware.SensorEventListener;
13   import android.hardware.SensorManager;
14   import android.os.Bundle;
15   import android.view.SurfaceHolder;
16   import android.view.SurfaceView;
17   import android.view.Window;
18   import android.view.WindowManager;
19   import android.view.SurfaceHolder.Callback;
20   public class MainActivity extends Activity
21   {
22    BallView mAnimView = null;
23    @Override
24    public void onCreate(Bundle savedInstanceState)
25    {
26      super.onCreate(savedInstanceState);
27      /* 全屏显示窗口 */
28      requestWindowFeature(Window.FEATURE_NO_TITLE);
29      getWindow().setFlags(WindowManager.LayoutParams.FLAG_FULLSCREEN,
30      WindowManager.LayoutParams.FLAG_FULLSCREEN);
31      /* 强制横屏 */
32      setRequestedOrientation(ActivityInfo.SCREEN_ORIENTATION_LANDSCAPE);
33      /* 显示自定义的游戏View */
34      mAnimView = new BallView(this);
35      setContentView(mAnimView);
36    }
37
38   public class BallView extends SurfaceView
39         implements Callback,Runnable ,SensorEventListener
40   {
41     /** 每50帧刷新一次屏幕 **/
42     public static final int TIME_IN_FRAME = 50;
43     /** 游戏画笔 **/
44     Paint mPaint = null;
45     Paint mTextPaint = null;
46     SurfaceHolder mSurfaceHolder = null;
47     /** 控制游戏更新循环 **/
48     boolean mRunning = false;
49     /** 游戏画布 **/
50     Canvas mCanvas = null;
51     /** 控制游戏循环 **/
52     boolean mIsRunning = false;
53     /** SensorManager管理器 **/
```

```
54      private SensorManager mSensorMgr = null;
55      Sensor mSensor = null;
56      /** 手机屏幕的宽、高 **/
57      int mScreenWidth = 0;
58      int mScreenHeight = 0;
59      /** 小球资源文件越界区域 **/
60      private int mScreenBallWidth = 0;
61      private int mScreenBallHeight = 0;
62      /** 游戏背景文件 **/
63      private Bitmap mbitmapBg;
64      /** 小球资源文件 **/
65      private Bitmap mbitmapBall;
66      /** 小球的坐标位置 **/
67      private float mPosX = 200;
68      private float mPosY = 0;
69      /** 重力感应X轴、Y轴、Z轴的重力值 **/
70      private float mGX = 0;
71      private float mGY = 0;
72      private float mGZ = 0;
73      public BallView(Context context)
74      {
75          super(context);
76          /** 设置当前View拥有控制焦点 **/
77          this.setFocusable(true);
78          /** 设置当前View拥有触摸事件 **/
79          this.setFocusableInTouchMode(true);
80          /** 拿到SurfaceHolder对象 **/
81          mSurfaceHolder = this.getHolder();
82          /** 将mSurfaceHolder添加到Callback回调函数中 **/
83          mSurfaceHolder.addCallback(this);
84          /** 创建画布 **/
85          mCanvas = new Canvas();
86          /** 创建曲线画笔 **/
87          mPaint = new Paint();
88          mPaint.setColor(Color.WHITE);
89          /** 加载小球资源 **/
90          mbitmapBall = BitmapFactory.decodeResource(this.getResources(),
91                          R.drawable.ball);
92          /** 加载游戏背景 **/
93          mbitmapBg = BitmapFactory.decodeResource(this.getResources(),
94                          R.drawable.bg);
95          /** 获得SensorManager对象 **/
96          mSensorMgr = (SensorManager) getSystemService(SENSOR_SERVICE);
97          mSensor = mSensorMgr.getDefaultSensor(Sensor.TYPE_ACCELEROMETER);
98          mSensorMgr.registerListener(this, mSensor,
```

```
 99                                SensorManager.SENSOR_DELAY_GAME);
100     }
101     private void Draw()
102     {
103         /** 绘制游戏背景 **/
104         mCanvas.drawBitmap(mbitmapBg,0,0, mPaint);
105         /** 绘制小球 **/
106         mCanvas.drawBitmap(mbitmapBall, mPosX,mPosY, mPaint);
107         /** 显示X轴、Y轴、Z轴的重力值 **/
108         mCanvas.drawText("    X轴重力值 : " + mGX, 0, 20, mPaint);
109         mCanvas.drawText("    Y轴重力值 : " + mGY, 0, 40, mPaint);
110         mCanvas.drawText("    Z轴重力值 : " + mGZ, 0, 60, mPaint);
111     }
112     @Override
113     public void surfaceChanged(SurfaceHolder holder, int format,
114                         int width,int height) {      }
115     @Override
116     public void surfaceCreated(SurfaceHolder holder)
117     {
118         /** 开始游戏主循环线程 **/
119         mIsRunning = true;
120         new Thread(this).start();
121         /** 得到当前屏幕的宽、高 **/
122         mScreenWidth = this.getWidth();
123         mScreenHeight = this.getHeight();
124         /** 得到小球越界区域 **/
125         mScreenBallWidth = mScreenWidth - mbitmapBall.getWidth();
126         mScreenBallHeight = mScreenHeight - mbitmapBall.getHeight();
127     }
128     @Override
129     public void surfaceDestroyed(SurfaceHolder holder)
130     {
131         mIsRunning = false;
132     }
133     @Override
134     public void run()
135     {
136       while (mIsRunning)
137       {
138       /** 取得更新游戏之前的时间 **/
139       long startTime = System.currentTimeMillis();
140       /** 在这里加上线程安全锁 **/
141       synchronized (mSurfaceHolder)
142       {
143        /** 拿到当前画布,然后锁定 **/
```

```
144        mCanvas = mSurfaceHolder.lockCanvas();
145        Draw();
146        /** 绘制结束后解锁显示在屏幕上 **/
147        mSurfaceHolder.unlockCanvasAndPost(mCanvas);
148     }
149     /** 取得更新游戏结束的时间 **/
150     long endTime = System.currentTimeMillis();
151     /** 计算出游戏一次更新的毫秒数 **/
152     int diffTime = (int) (endTime - startTime);
153     /** 确保每次更新时间为50帧 **/
154     while (diffTime <= TIME_IN_FRAME)
155     {
156        diffTime = (int) (System.currentTimeMillis() - startTime);
157        /** 线程等待 **/
158        Thread.yield();
159     }
160   }
161 }
162 @Override
163 public void onAccuracyChanged(Sensor arg0, int arg1)
164 { }
165 @Override
166 public void onSensorChanged(SensorEvent event)
167 {
168   mGX = event.values[0];
169   mGY = event.values[1];     ← 获取 X、Y、Z 3 个方向重力加速度的改变值
170   mGZ = event.values[2];
171   mPosX += mGX * 2;          ← 这里乘 2 是为了让小球移动更快
172   mPosY += mGY * 2;
173   if (mPosX < 0)
174   {
175      mPosX = 0;
176   } else if (mPosX > mScreenBallWidth)
177   {
178      mPosX = mScreenBallWidth;
179   }
180   if (mPosY < 0)                          ← 检测小球是否超出边界
181   {
182      mPosY = 0;
183   } else if (mPosY > mScreenBallHeight)
184   {
185      mPosY = mScreenBallHeight;
186   }
187  }
188 }
```

```
189 }
```

该程序在真实手机上的运行结果如图 9.13 所示。

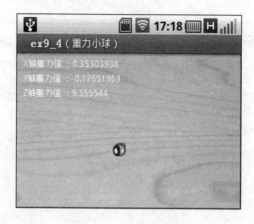

图 9.13 重力小球游戏

习 题 9

1. 编写一个地图应用程序，设已知 3 个地点的经纬度分别为（24.477384, 118.158216）、（24.488967, 118.144277）、（24.491091, 118.136781），在地图上画出它们的路线。

提示：

设

```
int x1, y1, x2, y2, x3, y3;
x1=(int)(24.477384 * 1000000);
y1=(int)(118.158216 * 1000000);
x2=(int)(24.488967 * 1000000);
y2=(int)(118.144277 * 1000000);
x3=(int)(24.491091 * 1000000);
y3=(int)(118.136781 * 1000000);
```

则

```
GeoPoint gpoint1, gpoint2, gpoint3;   //连线的点
gpoint1 = new GeoPoint(x1,y1);
gpoint2 = new GeoPoint(x2,y2);
gpoint3 = new GeoPoint(x3,y3);
mMapView.getController().animateTo(gpoint1);
```

2. 编写一个通过"摇一摇"播放音乐的音频播放器。
3. 进一步完善例 9-4，添加开始、结束等功能，使之成为一个较完整的简单游戏。

图书资源支持

感谢您一直以来对清华版图书的支持和爱护。为了配合本书的使用,本书提供配套的素材,有需求的用户请到清华大学出版社主页(http://www.tup.com.cn)上查询和下载,也可以拨打电话或发送电子邮件咨询。

如果您在使用本书的过程中遇到了什么问题,或者有相关图书出版计划,也请您发邮件告诉我们,以便我们更好地为您服务。

我们的联系方式:

地　　址: 北京海淀区双清路学研大厦 A 座 707

邮　　编: 100084

电　　话: 010-62770175-4604

资源下载: http://www.tup.com.cn

电子邮件: weijj@tup.tsinghua.edu.cn

QQ: 883604(请写明您的单位和姓名)

用微信扫一扫右边的二维码,即可关注清华大学出版社公众号"书圈"。

扫一扫
资源下载、样书申请
新书推荐、技术交流